MARKOV MODELS WITH COVARIATE DEPENDENCE FOR REPEATED MEASURES

MARKOV MODELS WITH COVARIATE DEPENDENCE FOR REPEATED MEASURES

M. ATAHARUL ISLAM
RAFIQUL ISLAM CHOWDHURY
AND
SHAHARIAR HUDA

Nova Science Publishers, Inc.
New York

For permission to use material from this book please contact us:
Telephone 631-231-7269; Fax 631-231-8175
Web Site: http://www.novapublishers.com

LIBRARY OF CONGRESS CATALOGING-IN-PUBLICATION DATA

Islam, M. Ataharul, 1976-
Markov models with covariate dependence for repeated measures / M. Ataharul Islam, Rafiqul Islam Chowdhury.
p. cm.
ISBN 978-1-60456-977-3 (hardcover)
1. Multivariate analysis. 2. Markov processes. I. Chowdhury, Rafiqul Islam, 1974- II. Title.
QA278.I75 2008
519.2'33--dc22
2008034444

Published by Nova Science Publishers, Inc. ✝ New York

CONTENTS

PREFACE

In recent years, there has been a growing interest in the longitudinal data analysis techniques. The longitudinal analysis covers a wide range of potential areas of applications in the fields of survival analysis and other biomedical applications, epidemiology, reliability and other engineering applications, agricultural statistics, environment, meteorology, biological sciences, econometric analysis, time series analysis, social sciences, demography, etc. In all these fields, the problem of analyzing adequately the data from repeated measures poses formidable challenge to the users and researchers. The longitudinal data is comprised of repeated measures on both outcome or response variables and independent variables or covariates. In the past, some important developments have provided ground for analyzing such data. The developments of the generalized linear models, the generalized estimating equations, multistate models based on proportional or nonproportional hazards, Markov chain based models, and transitional models, etc. are noteworthy. In some cases, attempts were also made to link the time series approaches to analysis of repeated measures data.

At this backdrop, we observe that there is still a great demand for clear understanding of the models for repeated measures in the context of the first or higher order Markov chain. More importantly, until now there is not much available literature in modeling the repeated measures data linking the Markov chains with underlying covariates or risk factors. Whatever little has been published is scattered over various specialized journals that researchers and users from other fields may find difficulty in accessing. In other words, there is a serious lack of books on the covariate dependent Markov models where transition probabilities can be explained in terms of the underlying factors of interest. This book provides in a single volume, a systematic illustration of the development of the covariate dependent Markov models. The estimation and test procedures are also discussed with examples from the real life. Outlines of the computer programs used for these examples are also provided with brief illustrations. The detailed programs will be provided on request.

This book is suitable for both the users of longitudinal data analysis as well as for researchers in various fields. Although the examples provided are from the health sciences, similar examples could be obtained from all the disciplines we have mentioned earlier without changing the underlying theory. The applications are provided in details along with the theoretical background for employing such models so that the users can apply the models independently on the basis of the theory and applications provided in the book. Both statisticians and users of statistics with some background in using longitudinal data analysis problems will find the approach easily comprehensible.

This book contains twelve chapters and includes an appendix with the guideline for computer programming for each chapter. The chapters are organized as follows:

Chapter 1 provides a brief background and description of some data. The set of data used in this book for applications of various models extensively is a public domain data set which can be downloaded from the website after obtaining necessary permission.

Chapter 2 includes some preliminaries on probability and Markov chains which are necessary to understand the theoretical exposition outlined in the book. The necessary background materials are presented in a simple manner for a wide range of potential users including those with little knowledge of statistics.

Chapter 3 provides a background discussion on the generalized linear models and the logistic regression model. The logistic regression models for binary or polytomous outcomes are used quite extensively in this book. Chapter 3 exposition will help the readers to comprehend the later chapters easily.

Chapter 4 presents the theory and applications of the two state first order Markov model with covariate dependence. The exposition of the model is provided in a simple manner so that all the users can be familiar with both the theory and applications without much effort.

Chapter 5 is an extension of the two state first order covariate dependent Markov model discussed in Chapter 4. This chapter acts as a link between Chapter 4 and Chapter 6. This chapter introduces the readers to the two state second order covariate dependent Markov model.

Chapter 6 generalizes the two state covariate dependent Markov models to any order. The estimation and test procedures are highlighted and the models are illustrated with a data set for the third and fourth orders.

Chapter 7 introduces the multi-state covariate dependent first order Markov models. This is a generalization of Chapter 4 for any number of states. This chapter provides the necessary estimation and test procedures for any number of states with applications.

Chapter 8 is a further generalization of Chapter 6 and Chapter 7. Chapter 6 deals with higher order for two states and Chapter 7 introduces any number of states for the first order while Chapter 8 includes both the multistate as well as higher order. This chapter involves a large number of parameters hence the estimation and test procedures become a little tedious.

Chapter 9 provides the theoretical aspect to deal with the likelihood function based on the repeated transitions where any state might be occupied for several follow-up times. A simplification in handling the transitions, reverse transitions and repeated transitions is highlighted in this chapter. The applications of the proposed model are also included in this chapter.

Chapter 10 summarizes some of the inferential procedures for the models, parameters, order of the models, serial dependence, and alternative procedures are described with applications. This chapter provides helpful insights to the readers regarding various decision making procedures based on the covariate dependent Markov models of the first or higher orders.

Chapter 11 displays the generalized linear model formulation of the higher order covariate dependent Markov models primarily with log link function. This chapter illustrates the suitability of log linear models in fitting the higher order Markov models with covariate dependence.

Chapter 12 presents some marginal and conditional models. The generalized estimating equations are also discussed. Both the marginal and conditional models are compared and the applications highlight their differences as well.

Chapter 1

REPEATED MEASURES DATA

1.0 INTRODUCTION

The study of longitudinal data has gained importance increasingly over time due to the advantage of such models in explaining the problems more comprehensively. In other words, longitudinal analysis provides age, cohort and period effects. On the other hand, the cross sectional studies deal with only single measures at a particular point in time. Hence it becomes difficult to provide any realistic explanation of age, cohort and period effects on the basis of cross sectional studies. Sometimes, such questions are examined by employing cross sectional data with very restrictive assumptions. In a longitudinal study, unlike in a cross sectional study, we observe repeated measures at different times within a specified study period. We can observe both the outcome and explanatory variables at different times. This provides the opportunity to examine the relationship between the outcome and explanatory variables over time in terms of the changes in the status of the outcome variables. This also poses a formidable difficulty in developing appropriate models for analyzing longitudinal data mainly due to correlation among the outcomes on the same individual/item at different times as well as due to formulation of a comprehensive model capturing the huge information generated by transitions during the period of study.

1.1 BACKGROUND

The Markov chain models are now quite familiar in various disciplines. In a time series data, for instance, we may have to assume that the current outcome depends only on the previous outcome, irrespective of the presence of a long series. This provides an example of first order Markovian assumption. This can be generalized to other disciplines. For example, if we consider disease status of an individual at a time t, then it would be logical to assume that the outcome depends on the status at the previous time, t-1. In a share market, the price of a share at time t may depend on the price at previous time, t-1. In case of meteorological problem of rainfall, we may assume that the status regarding rainfall depends on the status on the previous day. There are similar examples from other fields ranging from survival analysis/reliability to environmental problems, covering a wide range of potential applications. However, if we want to examine the relationships between transitions from one state to another with the potential risk factors, then we need to link regression models with the transition probabilities. This book will address the background and relevant statistical procedures for dealing with covariate dependence of transition probabilities. These models can be called transition models, in general terms.

The transition models appear to be naturally applicable to data generated from longitudinal studies. In recent times, there has been a growing interest in the Markov models. In the past, most of the works on Markov models dealt with estimation of transition probabilities for first or higher orders. An inference procedure for stationary transition probabilities involving k states was developed by Anderson and Goodman (1957). The higher order probability chains were discussed by Hoel (1954). The higher order Markov chain models for discrete variate time series appear to be restricted due to over-parameterization and several attempts have been made to simplify the application. We observe that several approaches are prevailing in the theory and applications of Markov chain models. Based on the work of Pregram (1980), estimation of transition probabilities was addressed for higher order Markov models (Raftery, 1985; Raftery and Tavare, 1994; Berchtold and Raftery, 2002) which are known as the mixture transition distributions (MTDs). These can be used for modeling of high-order Markov chains for a finite state space. Similarly, analysis of sequences of ordinal data from a relapsing remitting of a disease can be modeled by Markov chain (Albert, 1994). Albert and Waclawiw (1998) developed a class of quasi-likelihood models for a two state Markov chain with stationary transition probabilities for heterogeneous transitional data. However, these models deal with only estimation of transition probabilities.

Regier (1968) proposed a model for estimating odds ratio from a two state transition matrix. A grouped data version of the proportional hazards regression model for estimating computationally feasible estimators of the relative risk function was proposed by Prentice and Gloeckler (1978). The role of previous state as a covariate was examined by Korn and Whittemore (1979). Wu and Ware (1979) proposed a model which included accumulation of covariate information as time passes before the event and considered occurrence or non-occurrence of the event under study during each interval of follow up as the dependent variable. The method could be used with any regression function such as the multiple logistic regression model. Kalbfleisch and Lawless (1985) proposed other models for continuous time. They presented procedures for obtaining estimates for transition intensity parameters in homogeneous models. For a first order Markov model, they introduced a model for covariate dependence of log-linear type. None of these models could be generalized to higher order due to complexity in the formulation of the underlying models.

Another class of models has emerged for analyzing transition models with serial dependence of the first or higher orders on the basis of the marginal mean regression structure models. Azzalini (1994) introduced a stochastic model, more specifically, a first order Markov model, to examine the influence of time-dependent covariates on the marginal distribution of the binary outcome variables in serially correlated binary data. The Markov chains are expressed in transitional form rather than marginally and the solutions are obtained such that covariates relate only to the mean value of the process, independent of association parameters. Following Azzalini (1994), Heagerty and Zeger (2000) presented a class of marginalized transition models (MTMs) and Heagerty (2002) proposed a class of generalized MTMs to allow serial dependence of first or higher order. These models are computationally tedious and the form of serial dependence is quite restricted. If the regression parameters are strongly influenced by inaccurate modeling for serial correlation then the MTMs can result in misleading conclusions. Heagerty (2002) provided derivatives for score and information computations. Lindsey and Lambert (1998) examined some important theoretical aspects concerning the use of marginal models and demonstrated that there are serious limitations such as: (i) produce profile curves that do not represent any possible individual, (ii) show that a treatment is better on average when, in reality, it is poorer for each individual subject, (iii) generate complex and implausible physiological explanations with underdispersion in subgroups and problems associated with no possible probabilistic data generating mechanism.

In recent years, there has been a great deal of interest in the development of multivariate models based on the Markov Chains. These models have wide range of applications in the fields of reliability, economics, survival analysis, engineering, social sciences, environmental studies, biological sciences, etc. Muenz and Rubinstein (1985) employed logistic regression models to analyze the transition probabilities from one state to another but still there is a serious lack of general methodology for analyzing transition probabilities of higher order Markov models. In a higher order Markov model, we can examine some inevitable characteristics that may be revealed from the analysis of transitions, reverse transitions and repeated transitions. Islam and Chowdhury (2006) extended the model to higher order Markov model with covariate dependence for binary outcomes. It is noteworthy that the covariate dependent higher order Markov models can be used to identify the underlying factors associated with such transitions. In this book, it is aimed to provide a comprehensive covariate-dependent Markov Model for higher order. The proposed model is a further generalization of the models suggested by Muenz and Rubinstein (1985) and Islam and Chowdhury (2006) in dealing with event history data. Lindsey and Lambert (1998) observed that the advantage of longitudinal repeated measures is that one can see how individual responses change over time. They also concluded that this must generally be conditional upon the past history of a subject, in contrast to marginal analyses that concentrate on the marginal aspects of models discarding important information, or not using it efficiently. The proposed model is based on conditional approach and uses the event history efficiently. Furthermore, using the Chapman-Kolmogorov equations, the proposed model introduces an improvement over the previous methods in handling runs of events which is common in longitudinal data.

1.2 Data Description

In order to illustrate applications of the proposed models and methods we shall make repeated use of some of the longitudinal data sets in this book. Detailed descriptions of these data sets are provided here.

1.2.1 Health and Retirement Survey Data

A nationwide Longitudinal Study of Health, Retirement, and Aging (HRS) in the USA was conducted on individuals over age 50 and their spouses. The study was supported by the National Institute on Aging (NIA U01AG009740) and was administered by the Institute for Social Research (ISR) at the University of Michigan. Its main goal was to provide panel data that enable research and analysis in support of policies on retirement, health insurance, saving, and economic well-being. The survey elicits information about demographics, income, assets, health, cognition, family structure and connections, health care utilization and costs, housing, job status and history, expectations, and insurance. The HRS data products are available without cost to researchers and analysts. The interested readers can visit the HRS website (http://hrsonline.isr.umich.edu/) for more details about this data set. Respondents in the initial HRS cohort were those who were born during 1931 to 1941. This cohort was first interviewed in 1992 and subsequently every two years. A total of 12,652 respondents were included in this cohort. The panel data doccumented by the RAND, from the HRS cohort of seven rounds of the study conducted in 1992 (Wave 1), 1994 (Wave 2), 1996 (Wave 3), 1998 (Wave 4), 2000 (Wave 5), 2002 (Wave 6) and 2004 (Wave 7) will be used for various applications. Table 1.1 shows the number of respondents at different waves.

Table 1.1. Number of Respondents at Different Waves

Wave	Respondents Status			
	Non Responses/Dead		Respondent alive	
	Number	Percentage	Number	Percentage
1	0	0	12652	100.0
2	1229	9.7	11423	90.3
3	1877	14.8	10775	85.2
4	2410	19.0	10242	81.0
5	3022	23.9	9630	76.1
6	3445	27.2	9207	72.8
7	3879	30.7	8773	69.3

The following variables can be considered from the HRS data set:

1.2.1.1 Dependent Variables

We have used only a few outcome variables of interest in this book for the sake of comparison across chapters in analyzing longitudinal data. We have included definitions of some potential outcome variables of interest to the likely users. There are many other variables which are not discussed in this section but can be used for further examination. We have provided examples from mental health, self reported health, self reported change in health status, functional changes in mobility index and activities of daily living index.

A. Mental Health Index

Mental health index was derived using a score on the Center for Epidemiologic Studies Depression (CESD) scale. The CESD score is the sum of eight indicators (ranges from 0 to

8). The negative indicators measure whether the respondent experienced the following sentiments all or most of the times: depression, everything is an effort, sleep is restless, felt alone, felt sad, and could not get going. The positive indicators measure whether the respondent felt happy and enjoyed life, all or most of the time. These two were reversed before adding in the score. The score ranges from 0 to 8.

B. Change in Self Reported Health

These variables measure the change in self reports of health categories excellent, very good, good, fair, and poor. The health categories are numbered from 1 (excellent) to 5 (poor), so that positive values of the change in self reported health denote deterioration. This measure is not available in the baseline wave.

C. Self Report of Health Change

The HRS also directly asks about changes in health. The responses may be much better (1) somewhat better (2), same (3), somewhat worse (4), and much worse (5). Higher values denote health deterioration. In Wave 1 for the HRS entry cohort, the change in health is relative to one year ago; in subsequent waves, the changes are relative to the previous interview, two years ago.

D. Functional Limitations Indices

The RAND HRS Data contains six primary functional limitation indices. Those indices were chosen for their comparability with studies that measure functional limitations. A variable was first derived that indicates if the respondent had difficulty performing a task (0=no difficulty; 1=difficulty). The exact question asked of the respondent varies slightly across the four survey waves. However, their measure of difficulty was defined to be comparable across waves. All indices are the sum of the number of difficulties a respondent has completing a particular set of tasks and uses a definition of difficulty that is comparable across waves. The score ranges from 0 to 5. Following two indices will be used as outcome variables.

D.1 Mobility Index: The five tasks included in the mobility index are walking several blocks, walking one block, walking across the room, climbing several flights of stairs, climbing one flight of stairs.

Table 1.2 shows first 21 lines from the data for four respondents from different waves. First column is patient id, second column is follow-up number, third column is the dependent variable, and fourth onward are the independent variables.

In Table 1.2, Mobility is a binary dependent variable. There can be dependent variables with multiple categories. As mentioned earlier the data set we used in the book is a public domain data set. We can not provide the data set to any third party according to the data use condition. However, interested researcher can obtain the data set after acquiring necessary permission from the Health and Retirement Study site (http://hrsonline.isr.umich.edu/).

D.2 Activities of Daily Living Index: Includes the five tasks bathing, eating, dressing, walking across a room, and getting in or out of bed.

Frequency and percentage distributions of the five dependent variables are presented in Table 1.3. For application, we need to define the states and will recode these variables, which we will explain in appropriate sections. We are providing some examples of data sets which can be used by the readers. In this book we will use mostly data set D1.

Table 1.2. Sample Data File for the SAS Program

CASEID	Wave	Mobility	AGE	GENDER	White	Black
1	1	0	54	1	1	0
1	2	1	56	1	1	0
2	1	1	57	0	1	0
2	2	1	59	0	1	0
2	3	1	62	0	1	0
2	4	1	63	0	1	0
2	5	1	65	0	1	0
3	1	0	56	1	1	0
3	2	0	58	1	1	0
3	3	0	60	1	1	0
3	4	0	62	1	1	0
3	5	0	64	1	1	0
3	6	0	66	1	1	0
3	7	1	68	1	1	0
4	1	0	54	0	1	0
4	2	0	55	0	1	0
4	3	1	57	0	1	0
4	4	0	59	0	1	0
4	5	0	61	0	1	0
4	6	0	63	0	1	0
4	7	0	65	0	1	0

1.2.1.2 Independent Variables

In this section, we introduce some of the background variables that can be employed in analyzing the longitudinal data. All of these will not be employed for the examples in the subsequent chapters. These are enlisted here to provide an idea about the data set being employed in the book.

Age at interview of the respondents (in months and years),
Gender (male=1, female=0),
Education (years of education, 0 (= none), 1, 2, ...,17+),
Ethnic group (1=White/Caucasian, 2=Black/African American, and 3=other),
Current Marital Status (1= Married, 2= Married but spouse absent, 3= Partnered, 4= Separated, 5= Divorce, 6= Separated/Divorced, 7= Widowed, 8= Never Married) (This variable has been recoded as Married/partnered=1 and rest as Single=0),
Religion (1=Protestant, 2=Catholic, 3=Jewish, 4= none/no preference, and 5=other),
Health behaviors: Physical Activity or Exercise (0=no, 1=yes).

Beginning in Wave 7, the single question about physical activity is replaced with three questions about physical activity, which offer the choice of vigorous, moderate or light physical activity occurring every day, more than once per week, once per week, one to three times per month, or never.

Table 1.3. Frequency Distribution of Dependent Variables for Wave 1 (Baseline)

Dependent variables	Frequency	Percentage
Mental Health Index		
0	7840	62.0
1	2331	18.4
2	1178	9.3
3	524	4.1
4	270	2.1
5	200	1.6
6	143	1.1
7	97	.8
Change in Self Reported Health		
1. Excellent	2807	22.2
2. Very good	3481	27.5
3. Good	3544	28.0
4. Fair	1807	14.3
5. Poor	1013	8.0

Table 1.3. (Continued)

Dependent variables	Frequency	Percentage
Self Report of Health Change		
1. Much better	714	5.6
2. Somewhat better	1276	10.1
3. Same	9072	71.7
4. Somewhat worse	1248	9.9
5. Much worse	341	2.7
Missing	1	0.0
Mobility Index		
0	9036	71.4
1	1784	14.1
2	885	7.0
3	443	3.5
4	323	2.6
5	170	1.3
Missing	11	0.1
Activities of Daily Living Index		
0	11987	94.7
1	408	3.2
2	142	1.1
3	64	.5
4	36	.3
5	13	.1
Missing	2	.1

Drinking habits (0=no, 1=yes),

Body Mass Index (BMI): is weight divided by the square of height (weight / height^2),

Total household income in US $ (respondent & spouse),

Number of living children,

Medical care utilization: Hospitalization in previous 12 months (0=no, 1=yes),

Medical care utilization: Doctor (0=no, 1=yes),

Medical care utilization: Home Care (0=no, 1=yes).

The frequency distribution of the selected independent variables for Wave 1 (base line) is presented in Table 1.4.

Table 1.4. Frequency Distribution of Independent Variables for Wave 1 (Baseline)

Independent variables	Frequency	Percentage
Age in years		
Gender		
1. Male	5868	46.4
0. Female	6784	53.6
Education		
0 (None)	83	.7
1	29	.2
2	63	.5
3	140	1.1
4	104	.8
5	145	1.1
6	262	2.1
7	209	1.7
8	643	5.1
9	513	4.1
10	778	6.1

Table 1.4. (Continued)

Independent variables	Frequency	Percentage
Education (Continued)		
11	727	5.7
12	4424	35.0
13	783	6.2
14	1128	8.9
15	409	3.2
16	1040	8.2
17+	1172	9.3
Ethnic group		
1.White/Caucasian	10075	79.6
2.Black/African American	2095	16.6
3.Other	482	3.8
Marital Status		
1. Married/partnered	10222	80.8
0. Single	2430	19.2
Religion	8234	65.1
1.Protestant	3464	27.4
2.Catholic	217	1.7
3.Jewish	602	4.8
4.None/no preference	107	.8
5.Other	8234	65.1
Missing	28	0.2
Physical Activity or Exercise		
0.no	10199	80.6
1.yes	2453	19.4

Table 1.4. (Continued)

Independent variables	Frequency	Percentage
Drinking		
0.no	4996	39.5
1.yes	7656	60.5
Body Mass Index (BMI)		
<25.00 (Normal)	4524	35.8
25.00-29.99 (Overweight)	5147	40.7
30.00+ (Obese)	2981	23.6
Medical care utilization		
Hospital stay in previous 12 months		
0.no	11197	88.5
1.yes	1443	11.4
Missing	12	0.1
Doctor		
0.no	2625	20.7
1.yes	9969	78.8
Missing	58	0.5

1.2.2 Longitudinal Data on Maternal Morbidity in Bangladesh

This data set is based on a follow-up survey on maternal morbidity in Bangladesh, conducted by the Bangladesh Institute of Research for Promotion of Essential and Reproductive Health and Technologies (BIRPERHT) and funded by Ford Foundation. The data were collected from November 1992 to December 1993 (Akhter et al., 1996). A total of 1020 pregnant women were included in the study (pregnancy less than 6 months). Those subjects were followed-up with an interval of an average of 1 month, through full-term pregnancy, delivery and till 90 days postpartum period or 90 days after any other pregnancy outcome. The information on socio-economic background, pregnancy-related care and practice, extent of morbidity during the index pregnancy, delivery and postpartum period, or abortions were collected. We have considered up to sixth antenatal follow-ups for 992 women who had information on antenatal period from first follow-up. Table 1.5 showed the number of women at different follow-ups.

Table 1.5. Number of Respondents at Different Follow-Ups

Follow-up	Number of women
1	992
2	917
3	771
4	594
5	370
6	148

We may consider a number of variables as potential outcome variables from the above data set.

Dependent variables

1) Occurrence of haemorrhage (no=0, yes=1),
2) Occurrence of convulsion or fits (no=0, yes=1),
3) Edema of hands and feet (no=0, yes=1),
4) Cough and fever for more than 3 days (no=0, yes=1).

Table 1.6. Frequency Distribution of Dependent Variables for Baseline

Dependent variables	Frequency	Percentage
Occurrence of haemorrhage		
No	901	90.8
Yes	91	9.2
Occurrence of convulsion or fits		
No	867	87.4
Yes	125	12.6
Edema of hands and feet		
No	841	84.8
Yes	151	15.2

Table 1.6. (Continued)

Dependent variables	**Frequency**	**Percentage**
Cough and fever more than 3 days		
No	726	73.2
Yes	266	26.8

In this book, some of the following independent variables will be used.

Independent variables

Age of women (< 20 years = 0, 20 or more = 1),
Age at marriage (less than or equal to 15 years = 0, more than 15 years = 1),
Education (no education = 0, primary education = 1, secondary or higher education = 2),
Whether current pregnancy was wanted (no = 0, yes = 1),
Number of previous pregnancies (no previous pregnancy = 0, 1-4 = 1, 5+ = 2),
Economic status (low = 0, medium = 1, high = 2),
Antenatal visit for check-up (no visit = 0, regular/irregular visit = 1).

The frequency distribution of selected variables for baseline is presented in Table 1.7.

Table 1.7. Frequency Distribution of Selected Variables for Baseline

Dependent variables	**Frequency**	**Percentage**
Age of women		
less than 20 years	328	33.1
20 years or more	664	66.9
Age at marriage		
less than or equal to 15 years	460	46.4
more than 15 years	532	53.6
Education		
no education	544	54.8
primary	282	28.4

Table 1.7 (Continued)

Secondary or higher education	166	16.7
Whether current pregnancy was wanted		
No		
Yes		
Number of previous pregnancies		
no previous pregnancy	268	27.0
1-4 previous pregnancy	567	57.2
5 or more previous pregnancy	157	15.8
Economic status		
Low	240	24.2
Medium	552	55.6
High	200	20.2
Antenatal visit		
no antenatal visit	576	58.1
regular or irregular antenatal visit	416	41.9

1.2.3 A Data Set on Rainfall

The rainfall data from three districts in Bangladesh, namely, Rajshahi, Dhaka and Chittagong will be used in some of the applications in this book. The duration considered in this study ranges from 1964 to 1990. These secondary data were collected from the Department of Meteorology, Government of Bangladesh. We have considered the months of June to October in our applications. The reason behind considering this period is that it is typically considered as the Monsoon season and major agricultural crops are produced during this period in Bangladesh. The daily rainfall is measured in mm and is converted as dummy variable (no rain (0) and rain (1)) for each day during the study period for the months June to October. Three covariates are considered: wind speed (nautical miles/hour), humidity (relative humidity in percentage), and daily maximum temperature (measured in Celsius scale). The status of rainfall during the study period is displayed in Table 1.8.

Table 1.8 Distribution of Rainfall Status in Three Areas

	Occurrence of rain		Total
	No rain	Rain	
Dhaka	1670	2461	4131
Chittagong	1694	2437	4131
Rajshahi	2196	1935	4131

1.3 SUMMARY

This chapter provides the background for the book and also introduces some data sources which can be used for illustration of application of the models introduced in the subsequent chapters. In the past, most of the works on the theory and applications dealt with the estimation and test of transition probabilities without covariate dependence. Only in the last three decades, some models have been proposed in the area of Markov chain models with covariate dependence. In addition, although it appears to be very useful in explaining the problems associated with longitudinal or time series data, the models for higher order covariate dependent Markov chain models did not get much attention in the past. It is noteworthy that in the class of covariate dependent models, we may consider marginal and conditional approaches. Both types of models are illustrated in this book. The use of generalized linear model concepts are also described in explaining the Markov models of first or higher orders. The test procedures were not adequately described for covariate dependent Markov models in the previous works. This book provides use of several test procedures.

Chapter 2

MARKOV CHAIN: SOME PRELIMINARIES

2.1 PRELIMINARIES

In this chapter, some preliminary concepts related to the Markov chains are discussed. As the stochastic processes involve a collection or a family of random variables rather than a single random variable, we need to introduce the single variable based concepts first and then the stochastic processes will be defined. It is worth noting that the concepts of probability are key elements in understanding the Markov chains. In addition, the basic properties of random variable, sample space, conditional and unconditional probability, distributions, etc. will be required.

Probability

A random experiment is an experiment or observation which can be performed (at least in principle) any number of times under the same relevant conditions such that its outcome can not be predicted with certainty. Some examples are: (i) an individuals disease status at a follow-up time, (ii) whether or not it will rain next day, (iii) number of accidents at a particular place during the next hour, (iv) number of telephone calls during the next hour, (v) whether there will be a head in a coin toss, etc.

An outcome s is a possible result of a random experiment. For the experiments listed above respectively, some examples of outcomes are: (i) an individual will have a certain disease at the next follow-up, (ii) it will not rain next day, (iii) there were five accidents in the particular place during the specified hour, (iv) there were 3 telephone calls during the specified hour, (v) the coin toss resulted in a tail, etc.

The sample space S is the collection of all possible outcomes of a random experiment. An element in S is called a sample point. It is noteworthy that each outcome of an experiment is a sample point. For the examples above, the sample spaces are respectively: (i) S={Not Diseased, Diseased}, (ii) S={No Rain, Rain}, (iii) S={0,1,2,.....}, (iv) S={0,1,2,.....}, (v) S={Head (H), Tail (T)}.

An event E is a collection of outcomes in S, i.e. a subset of S, $E = \{s_1, \ldots, s_n \mid s_i \in S\}$.

A random variable is a numerical function that maps outcomes onto the real line, i.e., it assigns numerical values to the outcomes of an experiment. Usually, we denote a random variable by an upper case letter (*X*, *Y*, *Z*, etc.) and an observed value of the random variable is denoted by a lower case letter (*x*, *y*, *z*, etc.). We may consider that *X* denotes the random function at the time of an experiment being conducted before observing the outcome of the experiment and the observed value of the random variable X is denoted by x.

There are two different types of random variables on the basis of interval and ratio scales:

1) Discrete, i.e., the random variable takes on a countable number of values. For example, the number of customer arrivals at a bank in some time interval.
2) Continuous, i.e., there is an uncountable number of values of the random variable. For example, the service time at the bank.

In addition, we may consider categorical variables defined for nominal and ordinal scales. If we consider a variable without having natural order then it is called nominal, for instance, sex (male, female), race (Asian, Caucasian, others), religion (Christianity, Islam, Buddhism, Others), etc. Similarly, if there is an order in the categories of variable then we call it ordinal such as level of education (no education, primary, secondary, college or above), economic status (poor, middle, rich), disease status (normal, mild, severe), etc.

A probability function is a function that uniquely maps the events into the set of real values in [0,1]. If the sample space is S then the probability of the event E in the sample space S is denoted by P(E) which satisfies the following conditions:

(i) $0 \leq P(E) \leq 1$, (ii) P(S)=1, and (iii) for any sequence of mutually exclusive events

$$P\left(\cup_{n=1}^{\infty} E_n\right) = \sum_{n=1}^{\infty} P\left(E_n\right).$$

More specifically, let S be a sample space of a random experiment. A probability measure (function) P is an assignment of a real value P(E) to each event E⊂S that satisfies the following axioms:

1. P[Ø] = 0 and P[S] = 1,
2. 0≤ P[E] ≤ 1,
3. If $E_1 \cap E_2 = \emptyset$ (mutually exclusive) then $P(E_1 \cup E_2) = P(E_1) + P(E_2)$.

Other basic properties following the axioms are:

If $E_1 \subseteq E_2$ then P[E_2 - E_1] = P[E_2] - P[E_1], i.e., the probability of an outcome occurring which is in E_2 but not E_1 is P[E_2] - P[E_1].

$P[\bar{E}] = 1 - P[E]$ where $\bar{E}$ is the complement of E .

P[$E_1 \cup E_2$] = P[E_1] + P[E_2] - P[$E_1 \cap E_2$]

and, in general for n events ($E_1, E_2, \ldots, E_n$),

$$P\left[\bigcup_{i=1}^{n} E_i\right] = \sum_{i=1}^{n} P[E_i] - \sum_{i \neq j} P[E_i \cap E_j] + \sum_{i \neq j \neq k} P[E_i \cap E_j \cap E_k] + \cdots + (-1)^{n+1} P[\bigcap_{i=1}^{n} E_i] \cdot$$

It can be seen that

P[$E_1 \cup E_2$] ≤ P[E_1] + P[E_2] wh ch is known as Boole's Inequality.

If $E_1, E_2, \ldots, E_n$ are mutually exclusive events, then $P\left[\bigcup_{i=1}^{n} E_i\right] = \sum_{i=1}^{n} P[E_i]$.

For events E_1 and E_2, $E_1 \subseteq E_2$ implies $P[E_1] \leq P[E_2]$.

Conditional Probability

Let E_1 and E_2 be events in S. Then the conditional probability of E_1 given E_2 is

$P[E_1|E_2] = \dfrac{P[E_1 \cap E_2]}{P[E_2]}$, provided $P(E_2) \neq 0$, from which one obtains

$$P[E_1|E_2]P[E_2] = P[E_1 \cap E_2]$$

and $P[E_2|E_1] = \dfrac{P[E_1 \cap E_2]}{P[E_1]}$ thus getting $P[E_2|E_1] = \dfrac{P[E_1|E_2]P[E_2]}{P[E_1]}$.

Law of Total Probability

Let $E_1, E_2, \ldots, E_n$ be mutually exclusive and exhaustive events such that $\bigcup_{i=1}^{n} E_i = S$ i.e. $E_1, E_2, \ldots, E_n$ form an exhaustive partition of S. We know that due to mutual exclusivity, $E_i \cap E_j = \varnothing$ for all $i \neq j$. Let E be some event ⊆ S. Then

$$P[E] = \sum_{i=1}^{n} P[E|E_i]P[E_i].$$

Independent Events

Two events E_1 and E_2 are independent if and only if

$$P(E_1 \cap E_2) = P(E_1)P(E_2)$$

and, in terms of conditional probabilities, E_1 and E_2 are independent if and only if

$$P(E_1|E_2) = P(E_1), \quad \text{and } P(E_2|E_1) = P(E_2).$$

This can be generalized for n events i.e., $E_1, E_2, \ldots, E_n$ are independent if and only if

$$P(E_1 \cap E_2 \cap \ldots \cap E_n) = P(E_1)P(E_2)\ldots P(E_n).$$

Bayes Rule

Let $E_1, \cdots, E_n$ form a partition of S and let E be any event in S. Then

$$P[E_i \mid E] = \frac{P[E_i \cap E]}{P[E]}.$$

Using the law of total probability

$$P[E] = \sum_{j=1}^{n} P[E \mid E_j] P[E_j].$$

Substituting the above into the preceding equation yields

$$P[E_i \mid E] = \frac{P[E_i \cap E]}{\sum_{j=1}^{n} P[E \mid E_j] P[E_j]}$$

and use of $P[E_i \cap E] = P[E \cap E_i] = P[E \mid E_i] P[E_i]$ leads to

$$P[E_i \mid E] = \frac{P[E \mid E_i] P[E_i]}{\sum_{j=1}^{n} P[E \mid E_j] P[E_j]}$$

which is known as Bayes Rule.

Discrete Random Variables

A discrete random variable is one whose outcome values are from a countable set. Every discrete random variable X has a Probability Mass Function (pmf) denoted by $P(X = x) = p(x)$. Here, if the possible values of X are $x_1, x_2, \ldots$ then the discrete probabilities are

$P(X = x_i) = p(x_i) > 0$, i=1,2,... and $P(X = x) = p(x) = 0$ for all other values of x. According to the law of total probability

$$\sum_{i=1}^{\infty} P(X = x_i) = \sum_{i=1}^{\infty} p(x_i) = 1.$$

The cumulative distribution function (cdf) or distribution function of X is defined as

$$F(x) = P(X \leq x) = \sum_{x_i \leq x} p(x_i).$$

Continuous Random Variables

A continuous random variable is one whose outcome values are from an uncountable set. Every continuous random variable X has a Probability Density Function (pdf), *f(x)*, where

$$f(x) = \frac{d}{dx} F(x) \text{ and } F(x) = P(X \leq x) = \int_{-\infty}^{x} f(t)dt.$$

Expectation

The expected value of a random variable X is defined as

$$\mathrm{E[X]} = \sum_{x} \mathrm{x\ p}(x) \text{ if } X \text{ is a discrete random variable}$$

and

$$\mathrm{E[X]} = \int x f(x)\, dx \text{ if } X \text{ is a continuous random variable.}$$

Jointly Distributed Random Variables

We may be interested in systems involving more than one random variable. In case of two random variables, *X* and *Y*, we talk about their joint cumulative distribution function, i.e,

$$F(x, y) = P[X \leq x, Y \leq y].$$

Now, if X and Y are continuous, then

$$F(x,y)=\int_{-\infty}^{x}\int_{-\infty}^{y} f(u,v)du\,dv$$

and the joint density function of X and Y is given by

$$f(x,y)=\frac{\partial^2}{\partial x\partial y}F(x,y)\,.$$

However, if X and Y are discrete, then the joint probability mass function is

$$P(X=x,Y=y)=p(x,y)$$

and the joint cumulative distribution function is

$$F(x,y)=\sum_{s\leq x}\sum_{t\leq y} p(s,t)\,.$$

For continuous case marginal probability density functions for X and Y are, respectively,

$$f_X(x)=\int_{-\infty}^{\infty} f(x,y)dy,$$

$$f_Y(y)=\int_{-\infty}^{\infty} f(x,y)dx.$$

In discrete case the marginal probability mass functions for X and Y are, respectively,

$$p_X(x)=\sum_{y=-\infty}^{\infty} p(x,y),$$

$$p_Y(y)=\sum_{x=-\infty}^{\infty} p(x,y).$$

Two continuous random variables X and Y are independent if and only if

$$f(x,y)=f_X(x)f_Y(y) \text{ for every value of x and y}$$

where $f_X(x)$ and $f_Y(y)$ are the marginal pdf's of X and Y, respectively.

Similarly, in case of discrete random variables independence is equivalent to

$$p(x,y)=p_X(x)p_Y(y) \text{ for every value of x and y}$$

where $p_x(x)$ and $p_y(y)$ are the marginal pmf's of X and Y, respectively.

If X and Y are independent, then it can be shown that *E(XY)=E(X)E(Y)* for both discrete and continuous variables.

Conditional means and variances for continuous variables are defined in the usual way. The conditional mean and variance of Y given X = x are, respectively,

$$\mu_{Y|x} = E(Y|x) = \int_{-\infty}^{\infty} y f_y(y|x)dy,$$

$$\sigma^2_{Y|x} = Var(Y|x) = E[(Y - \mu_{Y|x})^2 |x] = \int_{-\infty}^{\infty} (y - \mu_{y|x})^2 f_y(y|x)dy,$$

where $f_y(y|x)$ is the conditional density Y given X=x.

Similarly, the conditional mean and variance of X given Y = y are, respectively,

$$\mu_{X|y} = E(X|y) = \int_{-\infty}^{\infty} x f_x(x|y)dx,$$

$$\sigma^2_{X|y} = Var(X|y) = E[(X - \mu_{X|y})^2 |y] = \int_{-\infty}^{\infty} (x - \mu_{x|y})^2 f_x(x|y)dx.$$

For discrete variables, the corresponding expressions are

$$\mu_{Y|x} = E(Y|x) = \sum_{y=-\infty}^{\infty} y p_y(y|x),$$

$$\sigma^2_{Y|x} = Var(Y|x) = E[(Y - u_{Y|x})^2 |x] = \sum_{y=-\infty}^{\infty} (y - \mu_{y|x})^2 p_y(y|x),$$

and

$$\mu_{X|y} = E(X|y) = \sum_{x=-\infty}^{\infty} x p_x(x|y),$$

$$\sigma^2_{X|y} = Var(X|y) = E[(X - \mu_{x|y})^2 |y] = \sum_{x=-\infty}^{\infty} (x - \mu_{x|y})^2 p_x(x|y).$$

If we consider n random variables $X_1, X_2, ..., X_n$ then the joint cumulative distribution function is

$$F(x_1,...,x_n) = P(X_1 \le x_1,..., X_n \le x_n) = \int_{-\infty}^{x_1} ... \int_{-\infty}^{x_n} f(z_1,..,z_n)dz_1...dz_n$$

for continuous variables, and

$$P[X_1 \le x_1,\ldots,X_n \le x_n] = \sum_{z_1=-\infty}^{x_1} \cdots \sum_{z_n=-\infty}^{x_n} p(z_1,\ldots,z_n)$$

for discrete variables where $f(x_1,\ldots,x_n)$ and $p(x_1,\ldots,x_n)$ are the joint probability density function and joint probability mass function, respectively. Then the expected values are:

$$\mu_i = E[X_i] = \int_{-\infty}^{\infty} \cdots \int_{-\infty}^{\infty} x_i f(x_1,\ldots,x_n) dx_1 \ldots dx_n$$

for continuous variables, and

$$\mu_i = E[X_i] = \sum_{x_1=-\infty}^{\infty} \cdots \sum_{x_n=-\infty}^{\infty} x_i p(x_1,\ldots,x_n)$$

for discrete variables.

2.2 Some Important Discrete Distributions

Bernoulli Distribution

A Bernoulli distribution is associated with an experiment consisting of a single trial where there can be one of two possible outcomes: success with probability p or failure with probability $q = (1\text{-}p)$. Let X be a Bernoulli random variable then

$$X = \begin{cases} 0 & \text{(failure)} \\ 1 & \text{(success)} \end{cases} \text{ and } p_X(k) = \begin{cases} 1-p & \text{if } k=0 \\ p & \text{if } k=1. \end{cases}$$

We can express the Bernoulli distribution as follows:

$$\begin{aligned} P(X=x) &= \binom{1}{x} p^x (1-p)^{1-x},\ (x=0,1)\ or \\ P(X=x) &= p^x (1-p)^{1-x},\ (x=0,1). \end{aligned} \tag{2.1}$$

From (2.1) it can be seen that E(X) = p and Var(X) = p(1-p).

Examples: Let us consider a single toss of a coin, the possible outcome being either head or tail. If the probability of head is p and the probability of tail is 1-p then it can be represented by the Bernoulli distribution with x=1 for head and x=0 for tail. Similarly, we may consider the operating state of a component (operating or failed) of a machine at some time. Here, we may consider the probability of failure p and probability of no failure as 1-p. Then with x=1 for failure and x=0 for no failure, we can represent it by the Bernoulli distribution described above.

Binomial Distribution

A Binomial distribution is associated with an experiment of *n* independent and identical Bernoulli trials with each trial resulting in a success with probability p or a failure with probability 1-p. Let *X* represent the number of successes. Then

$$p(X = x) = \binom{n}{x} p^x q^{n-x}, \quad x = 0,1,2,\ldots,n. \tag{2.2}$$

In this case, X is the sum of outcomes for n Bernoulli experiments.
From (2.2) it can be seen that E(X) = n p and Var (X) = n p (1-p).

Geometric Distribution

A Geometric distribution is associated with the number of trials in an experiment of independent Bernoulli trials that are performed until a success occurs. In this case X denotes the number of trials and

$$p_X(x) = pq^{x-1}, \quad x = 1,2,3,\ldots,\infty.$$

It can be seen that for a geometric distribution, the expected value and the variance are:

$$E(X) = (1-p)/p, \quad Var(X) = (1-p)/p^2.$$

Poisson Distribution

A Poisson distribution is associated with an experiment in which the number of occurrences of some outcome is counted over some time period t such that λ represents the rate of occurrences over the time period. In this case X is the number of occurrences and its probability mass function is

$$p(x, \lambda t) = \frac{(\lambda t)^x}{x!} e^{-\lambda t}, \quad x = 0,1,2,\ldots,\infty.$$

The Poisson distribution can be used to approximate the binomial distribution when *n* is large and *p* is small, substituting $\lambda t = np$ ($p = \lambda t / n$).

It can be seen that the expected value and variance of the Poisson distribution are

$$E(X) = \lambda t = Var(X).$$

An important characteristic of a Poisson distribution is the equality of mean and variance.

Multinomial Distribution

Binomial distribution arises when each trial has two possible outcomes. If there are more than two possible outcomes, then we consider multinomial distribution. Suppose that each of n independent, identical trials can have outcome in any of c categories.

Let $X_{ij} = 1$, if trial i has outcome in category j and $X_{ij} = 0$, otherwise.

Then $X_i = (X_{i1}, X_{i2}, ..., X_{ic})$ represents a multinomial trial with $\sum_{j=1}^{c} X_{ij} = 1$.

Let $n_j = \sum_{i=1}^{n} X_{ij}$ = number of trials having outcome in category j. Then the counts $(n_1, n_2, ..., n_c)$ have the multinomial distribution.

Let $p_j = P(Y_{ij} = 1)$ denote the probability of outcome in category j for each trial. The multinomial probability mass function is

$$p(n_1, n_2, ..., n_{c-1}) = \frac{n!}{n_1! n_2! ... n_c!} p_1^{n_1} p_2^{n_2} ... p_c^{n_c}.$$

Since $\sum_{i=1}^{c} n_i = n$, $n_c = n - (n_1 + n_2 + ... + n_{c-1})$ which shows that the dimension is (c-1) and thus $\sum_{j=1}^{c} p_j = 1$. For the multinomial distribution, it can be shown that

$$E(n_j) = np_j, \ \text{var}(n_j) = np_j(1 - p_j), \ \text{cov}(n_j, n_k) = -np_j p_k.$$

2.3 Two Important Continuous Distributions

Normal Distribution

The probability density function of a normally distributed random variable X, with mean $E(X) = \mu$ and variance $Var(X) = \sigma^2$, is given by the expression

$$f(x, \mu, \sigma^2) = \frac{1}{\sqrt{2\pi\sigma^2}} e^{-\frac{(x-\mu)^2}{2\sigma^2}} \qquad (-\infty < x < \infty).$$

(2.3)

The standardized normal variate is defined as

$$Z = \frac{X - \mu}{\sigma}$$

and the probability density function of the standardized normal variate Z is

$$f(z)=\frac{1}{\sqrt{2\pi}}e^{-\frac{z^2}{2}} \qquad (-\infty < Z < \infty). \tag{2.4}$$

The standardized normal variate, Z, is distributed as normal with mean 0 and variance 1 since (2.4) is a special case of (2.3) with $\mu = 0$ and $\sigma = 1$.

Chi-Square Distribution

If a continuous non-negative random variable U has the chi-square distribution with k degrees of freedom (χ_k^2) its probability density function is given by

$$f(u)=\frac{1}{\{(k/2)-1\}!2^{k/2}}u^{(k/2)-1}e^{-(u/2)}, \ (u>0).$$

Further, E(U) = k and Var(U) = 2k.

It is noteworthy that for a sample of n independent observations, X_i, i=1,2,…,n, from a normal population we can define

$$\chi_n^2 = \sum_{i=1}^{n}\frac{(X_i-\mu)^2}{\sigma^2} = \sum_{i=1}^{n} Z_i^2 .$$

and this follows a χ^2 distribution with degrees of freedom n.

2.4 Markov Chains

A stochastic process is a collection of random variables $\{X(t), t \in T\}$ indexed by a parameter such as time (or space). For a given t, $X(t)$ is a random variable referred to as the state of the system at time t. The values assumed by $X(t)$ are called states and the set of possible values is called the state space. The set of possible values of the indexing parameter t, i.e. T is called the index set or parameter space.

Some example of the state of a system at time t are: (i) rainfall or no rainfall on day t, (ii) disease or no disease on day t, (iii) functional ability of movement of an elderly on day t, (iv) accident or no accident on particular street corner at time t, (v) incidence of psychiatric problems during month t, etc. These are examples with only two states which can be denoted by responses "yes" or "no". However, we can extend the state space to contain more than two states, such as severity of disease, comprising of three states, normal, mild or severe state of the disease. Similarly, there can be state space with any number of states.

A stochastic process with a countable index set is known as a discrete time stochastic process. In this case we let $t = 0,1,2,\ldots$ and denote the states as $X_0, X_1, X_2, \ldots$. As an example, for rainfall we may consider an index set $T = \{0,1,2,\ldots,365\}$ for a year. If the index set is an interval, then the stochastic process is called a continuous time stochastic process. For example we may consider rainfall at a time t within a day in which case $X(t)$ denotes whether it is raining or not at time t and T denotes the 24 hour period.

Since a stochastic process is a collection of random variables, we need to consider joint distribution of $\{X(t), t \in T\}$. A stochastic process is called time homogeneous if $F(x_0, x; t_0, t_0 + t) = F(x_0, x; 0, t)$ for $t_0 \in T$, where $F(x, y; t_1, t_2)$ is the joint cumulative distribution function of $X(t_1)$ and $X(t_2)$ i.e., $F(x, y; t_1, t_2) = P(X(t_1) \le x, X(t_2) \le y)$. A stochastic process is called a Markov process if it has the property that

$$P(X(t) \le x \mid X(t_i) \le x_i; i = 0,1,...,n) = P(X(t) \le x \mid X(t_n) \le x_n)$$

A Markov chain is a Markov process whose state space is a finite or countable set, and whose (time) index set is $T = (0,1,2,...)$. This can be formally expressed as follows:

$$P(X_{n+1} = j \mid X_0 = i_0, ..., X_{n-1} = i_{n-1}, X_n = i) = P(X_{n+1} = j \mid X_n = i) \quad (2.5)$$

for all time points and for all states.

Here, we have considered a one-step transition probability from time point n to time point n+1. The one-step transition probability on the right hand side of (2.5) can be written as:

$$P_{ij}^{n,n+1} = P(X_{n+1} = j \mid X_n = i). \quad (2.6)$$

Let us consider here two states for X(t) = 0,1 and two time points for T= {0,1}. Then using (2.6) the transition probability matrix can be defined as follows:

$$P = \begin{pmatrix} P_{00}^{01} & P_{01}^{01} \\ P_{10}^{01} & P_{11}^{01} \end{pmatrix}. \quad (2.7)$$

Usually, we express this matrix of 1-step transition probabilities in the following form:

$$P = \begin{pmatrix} P_{00} & P_{01} \\ P_{10} & P_{11} \end{pmatrix}. \quad (2.8)$$

Similarly, (2.7) can be extended for three states X(t) = 0,1,2 and two time points T = {0,1} as follows:

$$P = \begin{pmatrix} P_{00}^{01} & P_{01}^{01} & P_{02}^{01} \\ P_{10}^{01} & P_{11}^{01} & P_{12}^{01} \\ P_{20}^{01} & P_{21}^{01} & P_{22}^{01} \end{pmatrix} \quad (2.9)$$

and (2.9) can be expressed in the following usual notations:

$$P = \begin{pmatrix} P_{00} & P_{01} & P_{02} \\ P_{10} & P_{11} & P_{12} \\ P_{20} & P_{21} & P_{22} \end{pmatrix}, \tag{2.10}$$

in analogy with (2.8). The transition probability matrix (2.10) can be extended for any finite or infinite time and state spaces. In general, we express the transition probability matrix as $P = \left\{P_{ij}\right\}$. The transition probabilities satisfy the following conditions:

$$P_{ij} \geq 0, \ \ i, j = 0,1,2,....; \ \ \sum_{j=0}^{\infty} P_{ij} = 1, \ \ i = 0,1,2,....$$

To define a Markov process, we need to specify its transition probability matrix and probability of initial state, X_0.

Let $P(X_0 = i) = p_i$. Then it can be seen that

$$P\left(X_0 = i, X_1 = i_1, ..., X_n = i_n\right) =$$
$$P(X_n = i_n | X_0 = i, X_1 = i_1, ..., X_{n-1} = i_{n-1}) \times P\left(X_0 = i, X_1 = i_1, ..., X_{n-1} = i_{n-1}\right)$$

which can be expressed as follows for Markov processes:

$$P\left(X_0 = i, X_1 = i_1, ..., X_n = i_n\right) =$$
$$P(X_n = i_n | X_{n-1} = i_{n-1}) \times P\left(X_0 = i, X_1 = i_1, ..., X_{n-1} = i_{n-1}\right) =$$
$$P_{i_{n-1}, i_n} \times P\left(X_0 = i, X_1 = i_1, ..., X_{n-1} = i_{n-1}\right).$$

Similarly,

$$P\left(X_0 = i, X_1 = i_1, ..., X_n = i_n\right) =$$
$$P_{i_{n-1}, i_n} \times P\left(X_0 = i, X_1 = i_1, ..., X_{n-1} = i_{n-1}\right) =$$
$$P_{i_{n-1}, i_n} \times P(X_{n-1} = i_{n-1} | X_{n-2} = i_{n-2}) \times P\left(X_0 = i, X_1 = i_1, ..., X_{n-2} = i_{n-2}\right) =$$
$$P_{i_{n-1}, i_n} \times P_{i_{n-2}, i_{n-1}} \times P\left(X_0 = i, X_1 = i_1, ..., X_{n-2} = i_{n-2}\right).$$

Repeating this for other states, we obtain the final expression in terms of transition probabilities:

$$P\left(X_0 = i, X_1 = i_1, ..., X_n = i_n\right) = P_{i_{n-1}, i_n} \times P_{i_{n-2}, i_{n-1}} \times ... \times P_{i_0, i_1} \times P_{i_0}. \tag{2.11}$$

It is also evident from the Markov property that the following relationship holds as well:

$$P\left(X_{n+1}=j_1,X_{n+2}=j_2,...,X_{n+m}=j_m\left|X_0=i_0,...,X_n=i_n\right.\right)=$$
$$P\left(X_{n+1}=j_1,X_{n+2}=j_2,...,X_{n+m}=j_m\left|X_n=i_n\right.\right). \quad (2.12)$$

Equations (2.11) and (2.12) illustrate two very desirable properties of a Markov chain.

n-step Transition Probabilities

The n-step transition probability of a Markov chain can be defined as

$$P_{ij}^{(n)}=P\left(X_{n+m}=j\left|X_m=i\right.\right)$$

which shows the probability of making a transition from state i to state j in n steps. If we assume that the Markov chain is homogeneous with respect to time, then the above probability is invariant of m. The n-step transition probabilities satisfy the following relationship:

$$P_{ij}^{(n)}=\sum_{k=0}^{\infty}P_{ik}P_{kj}^{(n-1)}, \quad (2.13)$$

$$P_{ij}^{(0)}=1 \text{ if } i=j, \text{ 0 otherwise.}$$

Equation (2.13) is equivalent to

$$P^{(n)}=P\times P\times ...\times P=P^n, \quad (2.14)$$

where $P^{(n)}$ is the matrix of n-step transition probabilities.

Based on the above property (2.14), we can describe the Chapman-Kolmogorov equation for computing $P^{(n)}$. The Chapman-Kolmogorov equation is defined as

$$P_{ij}^{(n+m)}=\sum_{k=0}^{\infty}P_{ik}^{(n)}P_{kj}^{(m)}.$$

(2.15)

In (2.15) the transition from i to j is observed in n + m steps, however, this transition is realized through an intermediate state k. In other words, a transition from i to k is observed in n steps and then the transition from k to j is realized in another m steps. In this case, we have to consider all the paths of intermediate transition to k and all the paths from k to j in order to obtain the desired probability for a transition from i to j in n + m steps.

Classification of States

Let us consider a s-state chain with states $0,1,...,s-1$ and the transition probability matrix

$$P = \begin{pmatrix} P_{01}...P_{1,s-1} \\ . \\ . \\ . \\ P_{s-1,1}...P_{s-1,s-1} \end{pmatrix} = (P_{ij}), \qquad 0 \le i, j \le s-1.$$

Let us define some of the important classification of states for a transition probability matrix. We can define a state j as accessible from state i if for some $n \ge 0$, $P_{ij}^{(n)} > 0$. To denote accessibility from i to j we use $i \to j$ and similarly, if $i \to j$ and $j \to i$ then i and j are intercommunicating states and we can denote it as $i \leftrightarrow j$. It is evident that, $P_{ii}^{(0)} = 1$ for all i in the state space.

Absorbing State

An absorbing state i is defined as a state such that if once a transition is made to the state, there is no other transition from that state. It is characterized by the following:

$$P_{ii} = 1, \qquad P_{ij} = 0, \quad i \ne j, \; j = 0,1,...,s-1.$$

Periodic State

The state, i, is called periodic if a t ($t > 1$) exists such that

$$P_{ii}^{(n)} = 0 \quad \text{for } n \ne t, 2t, 3t, ...$$

$$P_{ii}^{(n)} \ne 0 \quad \text{for } n = t, 2t, 3t, ...$$

where $P_{ii}^{(n)}$ is the probability of a return to state i. Here the state i is called periodic with period t. If a state does not satisfy this condition for periodicity then it is called aperiodic.

Persistent State

Let $f_i^{(n)}$ denote the probability of first return to state i and $P_{ii}^{(n)}$ denote the probability of a return (not necessarily the first return) to state i. Then $P_{ii}^{(n)}$ can be expressed as

$$P_{ii}^{(n)} = f_i^{(n)} + \sum_{r=1}^{n-1} f_i^{(r)} P_{ii}^{(n-r)}, \qquad n \ge 2.$$

The probability of a chain to return to the same state is obtained from

$$f_i = \sum_{n=1}^{\infty} f_i^{(n)}.$$

The state i is called a persistent state if $f_i = 1$. The mean recurrence time of a persistent state is

$$\mu_i = \sum_{n=1}^{\infty} n f_i^{(n)} .$$

Transient State

If there is uncertainty about returning to state i, that is, if

$$f_i = \sum_{n=1}^{\infty} f_i^{(n)} < 1$$

then state i is called transient.

Ergodic State

The state i is called ergodic if it is persistent, non-null and aperiodic.

Classification of Chains

Above we have discussed the classification of states. Next we provide some elementary concepts regarding classification of chains.

Irreducible Chains

If every state of a chain can be reached or is accessible from any other state in the same chain in a finite number of steps then it is defined as an irreducible chain. For any state i in an irreducible chain $P_{ii}^{(n)} > 0$ for some integer n. If for a transition probability matrix, P, there exists an integer r such that all elements of P^r are strictly positive then the transition probably matrix is said to be regular. A regular chain is irreducible.

Closed Sets

In a Markov chain, there may exist some subchains with some states being in the class of persistent, transient, absorbing, etc. These subchains can be defined as closed sets. Let us consider a set of states, E, within the Markov chain to be persistent if any state within E can be accessible from any other state in E, but can not be accessed from outside E to inside or vice versa. Then E is known as a closed persistent set. Similarly, we can define closed sets for transient states, absorbing states, etc.

Ergodic Chains

All states in an irreducible chain belong to the same class. If all states are ergodic, that is, persistent, non-null and aperiodic then the chain is known as an ergodic chain.

Some Important Limit Theorems

For any recurrent state i, the probability of first return is expressed as

$$f_i^{(n)} = P\left(X_n = i \middle| X_0 = i\right)$$

with

$$f_i = \sum_{n=1}^{\infty} f_i^{(n)} = 1$$

and mean time of recurrence to state i is

$$\mu_i = \sum_{n=1}^{\infty} n f_i^{(n)} .$$

Then we can also show for a recurrent, irreducible and aperiodic Markov chain that the limiting probability of returning to state i (not necessarily for the first time) is

$$\lim_{n\to\infty} P_{ii}^{(n)} = \frac{1}{\sum_{n=0}^{\infty} n f_i^{(n)}} = \frac{1}{\mu_i}.$$

It can also be shown for a positive recurrent aperiodic class with states j, j=0,1,2,...that

$$\lim_{n\to\infty} P_{jj}^{(n)} = \pi_j = \sum_{i=0}^{\infty} \pi_i P_{ij}$$

where π_i's are stationary probabilities of the Markov chain. By definition, the joint stationary probability for states i and j is

$$P\left(X_n = i, X_{n+1} = j\right) = \pi_i P_{ij}$$

and the limiting distribution for any state X_j is

$$\lim_{n\to\infty} P_{ij}^{(n)} = \pi_j = \lim_{n\to\infty} P\left(X_n = j \middle| X_0 = i\right).$$

2.5 Higher Order Markov Chains

Let us consider a single stationary process $(Y_{i1}, Y_{i2}, ..., Y_{ij})$ representing the past and present responses for subject i (i= 1,.2,...,n) at follow-up j (j=1,2,..., n_i). Y_{ij} is the response at time t_{ij}. We can think of Y_{ij} as an explicit function of past history of subject i at follow-up j denoted by $H_{ij} = \{Y_{ik}, \text{ k=1,2,...,j-1}\}$. The transition models for which the conditional distribution of Y_{ij} given H_{ij} depends on r prior observations, $Y_{ij-1}, ..., Y_{ij-r}$ is considered as the model of order r. For simplicity we first consider the case of binary response. The binary outcome is defined as y_{ij}=1, if an event occurs for the ith subject at the jth follow-up, Y_{ij}=0, otherwise. We replace p_{ij} by π_{ij} .

Then the first order Markov model can be expressed as

$$P(Y_{ij} | H_{ij}) = P(Y_{ij} | Y_{ij-1}) \tag{2.16}$$

and the transition probability matrix corresponding to (2.16) is given as Y_{ij}

$$\begin{array}{c|c} & Y_{ij-1} \quad 0 \qquad\quad 1 \\ \begin{matrix}0\\1\end{matrix} & \begin{bmatrix}\pi_{00} & \pi_{01}\\ \pi_{10} & \pi_{11}\end{bmatrix}. \end{array} \tag{2.17}$$

In (2.17), the probability of a transition from state 0 at time t_{j-1} to state 1 at time t_j is $\pi_{01} = P(Y_{ij} = 1 | Y_{ij-1} = 0)$ and similarly the probability of a transition from 1 at time t_{j-1} to 1 at time t_j is $\pi_{11} = P(Y_{ij} = 1 | Y_{ij-1} = 1)$.

The second order Markov model can be expressed as

$$P(Y_{ij} | H_{ij}) = P(Y_{ij} | Y_{ij-2}, Y_{ij-1}). \tag{2.18}$$

For second order Markov models Y_{ij} is a function of past history of subject i at follow-up j denoted by $H_{ij}^* = \{Y_{ik},\ \text{k=j-1,j-2}\}$. In other words, as (2.18) shows, the transition model for order 2 presents the conditional distribution of y_{ij} given H_{ij} depending on 2 prior observations Y_{ij-1}, Y_{ij-2} where Y_{ij-1}, Y_{ij-2} =0, 1.

The second order transition probabilities for time points t_{ij-2}, t_{ij-1} and t_{ij} at follow-up j with corresponding outcomes Y_{ij-2}, Y_{ij-1} and Y_{ij}, respectively, can be shown as follows:

Y_{ij-2}	Y_{ij-1}	Y_{ij} = 0	Y_{ij} = 1
0	0	π_{000}	π_{001}
0	1	π_{010}	π_{011}
1	0	π_{100}	π_{101}
1	1	π_{110}	π_{111}

. (2.19)

The transition models for which the conditional distribution of Y_{ij} given H_{ij} depends on r prior observations, $Y_{ij-1},...,Y_{ij-r}$, is considered as the model of order r.

Then the Markov chain of order r can be expressed as

$$P(Y_{ij} | H_{ij}) = P(Y_{ij} | Y_{ij-r},...,Y_{ij-1}). \tag{2.20}$$

Equation (2.20) generalizes (2.16) and (2.18). In other words, all the r immediate past observations are considered here and the conditional distribution of Y_{ij} given H_{ij} depending on r prior observations $Y_{ij-1}, Y_{ij-2},...,Y_{ij-r}$ where $Y_{ij-1}, Y_{ij-2},..., Y_{ij-r}$ =0, 1. It is assumed that Y_{ij} is a function of past history of subject i at follow-up j denoted by $H_{ij}^{*} = \{Y_{ik}, \text{ k=j-1,j-2,...,j-r}\}$ for Markov models of order r.

For order r, we need to consider 2^r sets of models. The transition probability matrix for the r-th order Markov model with outcomes with Y_{ij}=0 or 1, at time points $t_{i,j-r},...,t_{i,j-1},t_j$ are represented by $Y_{i,j-r},...,Y_{i,j-1},Y_j$ respectively. The outcomes are represented in the following matrix for the m-th transition type:

$Y_{i,j-r}$	$Y_{i,j-(r-1)}$	...	$Y_{i,j-1}$	Y_{ij}	
				0	1
0	0	...	0	0	1
0	0	...	1	0	1
$s_{m,i,j-r}$	$s_{m,i,j-(r-1)}$	...	$s_{m,i,j-1}$	$s_{m,i,j}$=0	$s_{m,i,j}$=1
.	.		.	.	.
.	.		.	.	.
..			.	.	.
1	1	...	1	0	1

(2.21)

where m denotes one of 2^r transition types generated from a Markovian transition probability matrix of order r. Note that (2.21) generalizes (2.19) which in turn is a generalization of (2.17).

The transition probabilities may be displayed as:

$Y_{i,j-r}$	$Y_{i,j-(r-1)}$	...	$Y_{i,j-1}$	Y_{ij}	
				0	1
0	0	...	0	$\pi_{0...00}$	$\pi_{0...01}$
0	0	...	1	$\pi_{0...11}$	$\pi_{0...11}$
.			.		.
.			.		.
.			.		.
$s_{m,i,j-r}$	$s_{m,i,j-(r-1)}$		$s_{m,i,j-1}$	$\pi_{s_{m,i,j-r},...,s_{m,i,j-1},0}$	$\pi_{s_{m,i,j-r},...,s_{m,i,j-1},1}$
.				.	
.				.	
.				.	
1	1	...	1	$\pi_{1...10}$	$\pi_{1...11}$

(2.22)

Sum of all the row probabilities are 1. Note that (2.22) displays transition probabilities for (2.21).

2.6 STATISTICAL INFERENCE

Two State Markov Chains

Let us consider the transition counts for a two state Markov chain as shown below:

$$\begin{array}{c|c} & Y_{ij} \\ Y_{ij-1} & \begin{array}{cc} 0 & 1 \end{array} \\ \begin{array}{c} 0 \\ 1 \end{array} & \begin{bmatrix} n_{00} & n_{01} \\ n_{10} & n_{11} \end{bmatrix}. \end{array} \tag{2.23}$$

Here the marginal totals are: $n_{00}+n_{01}=n_{0.}$, $n_{10}+n_{11}=n_{1.}$. The total number of observations is $n_{0.}+n_{1.}=n_{..}$.

The transition probability matrix, as stated in (2.17) is

$$\begin{array}{c|c} & Y_{ij} \\ Y_{ij-1} & \begin{array}{cc} 0 & 1 \end{array} \\ \begin{array}{c} 0 \\ 1 \end{array} & \begin{bmatrix} \pi_{00} & \pi_{01} \\ \pi_{10} & \pi_{11} \end{bmatrix}. \end{array}$$

From (2.17) and (2.23) the likelihood function is

$$L=\prod_{i=0}^{1}\binom{n_{i.}}{n_{i0}}\pi_{i0}^{n_{i0}}\pi_{i1}^{n_{i1}}. \tag{2.24}$$

From (2.24) the maximum likelihood estimates of the transition probabilities are

$$\hat{\pi}_{i0}=\frac{n_{i0}}{n_{i.}},\quad i=0,1. \tag{2.25}$$

Then using (2.25) the test statistic for testing the null hypotheses $H_0:\pi_{ij}=\pi_{ij}^0$ i=1,2; j=1,2 is

$$\chi^2=\sum_{i=0}^{1}\sum_{j=0}^{1}\frac{n_{i.}(\hat{\pi}_{ij}-\pi_{ij}^0)^2}{\pi_{ij}^0} \tag{2.26}$$

where the χ^2 in (2.26) is distributed as χ^2 with 2 degrees of freedom.

Similarly, the likelihood ratio test shows

$$-2\ln\Lambda=2\sum_{i=0}^{1}\sum_{j=0}^{1}n_{ij}\ln\frac{n_{ij}}{n_{i.}\pi_{ij}^0}, \tag{2.27}$$

where Λ is likelihood ratio. This (2.27) is χ^2 with 2 degrees of freedom under H_0. See Bhat (1971) for details.

2.7 S-STATE MARKOV CHAIN

For a s-state Markov chain, the transition counts are

$$\begin{array}{c|c} Y_{ij-1} & Y_{ij} \\ & \begin{array}{cccc} 0 & 1 & \cdots & s-1 \end{array} \\ \begin{array}{c} 0 \\ 1 \\ \cdot \\ \cdot \\ \cdot \\ s-1 \end{array} & \begin{bmatrix} n_{00} & n_{01} & \cdots & n_{0,s-1} \\ n_{10} & n_{11} & \cdots & n_{1,s-1} \\ \cdot & & & \\ \cdot & & & \\ \cdot & & & \\ n_{s-1,0} & n_{s-1,1} & \cdots & n_{s-1,s-1} \end{bmatrix} \end{array} . \tag{2.28}$$

Here the marginal totals are:

$n_{11}+\ldots+n_{1s}=n_{1.}$, $n_{21}+\ldots+n_{2s}=n_{2.},\ldots,n_{s1}+\ldots+n_{ss}=n_{s.}$. The total number of observations is $n_{1.}+n_{2.}+\ldots+n_{s.}=n_{..}$.

The transition probability matrix is

$$\begin{array}{c|c} Y_{ij-1} & Y_{ij} \\ & \begin{array}{cccc} 0 & 1 & \cdots & s-1 \end{array} \\ \begin{array}{c} 0 \\ 1 \\ \cdot \\ \cdot \\ \cdot \\ s-1 \end{array} & \begin{bmatrix} \pi_{00} & \pi_{01} & \cdots & \pi_{0,s-1} \\ \pi_{10} & \pi_{11} & \cdots & \pi_{1,s-1} \\ \cdot & \cdot & & \\ \cdot & \cdot & & \\ \cdot & & & \\ \pi_{s-1,0} & \pi_{s-1,1} & \cdots & \pi_{s-1,s-1} \end{bmatrix} \end{array} \tag{2.29}$$

From (2.28) and (2.29), the likelihood function is

$$L=\prod_{i=0}^{s-1} \frac{n_i\,!}{n_{i1}!\ldots n_{is}\,!}\pi_{i1}^{n_{i1}}\ldots\pi_{is}^{n_{is}} . \tag{2.30}$$

From (2.30), the maximum likelihood estimates of the transition probabilities are

$$\hat{\pi}_{ij} = \frac{n_{ij}}{n_{i.}}, \quad i, j = 0, ..., s-1. \tag{2.31}$$

Then using (2.31) the test statistic for testing the null hypotheses $H_0 : \pi_{ij} = \pi_{ij}^0$ is

$$\chi^2 = \sum_{i=0}^{s-1} \sum_{j=0}^{s-1} \frac{n_{i.}(\hat{\pi}_{ij} - \pi_{ij}^0)^2}{\pi_{ij}^0}. \tag{2.32}$$

This is χ^2 with s(s-1)-d degrees of freedom under H_0 where d is the number of zero π_{ij}^0's. Equation (2.32) is an extension of (2.26).

The above test procedure can be extended for higher order Markov chain. Let us suppose that the order of the Markov chain is r. Then the test statistic is

$$\chi^2 = \sum_{i=0}^{s^r-1} \sum_{j=0}^{s-1} \frac{n_{i.}(\hat{\pi}_{ij} - \pi_{ij}^0)^2}{\pi_{ij}^0}.$$

Here, i denotes the row of the transition probability matrix with transition type based on previous r transition outcomes prior to the current transition. The 2-state (s=2) transition type has been illustrated in Section 2.5.

Similarly, generalizing (2.27) the likelihood ratio test for first order is

$$-2\ln\Lambda = 2\sum_{i=0}^{s-1} \sum_{j=0}^{s-1} n_{ij} \ln\frac{n_{ij}}{n_{i.}\pi_{ij}^0}, \tag{2.33}$$

where Λ is likelihood ratio. This is χ^2 with s(s-1) degrees of freedom under H_0. In this case also, we can extend the test (2.33) for order r as

$$-2\ln\Lambda = 2\sum_{i=0}^{s^r-1} \sum_{j=0}^{s-1} n_{ij} \ln\frac{n_{ij}}{n_{i.}\pi_{ij}^0},$$

which is χ^2 with $s^r(s-1)$ degrees of freedom under H_0.

2.8 STATIONARITY TEST

Let us suppose that for a s-state Markov chain, the transition counts are observed at T follow-ups for a one step transition defined by $\pi_{ij}^t = P[Y(t=1) = j | Y(t) = i]$, and the counts are shown below for s states

t j

$$\begin{array}{c} 1 \\ 2 \\ .. \\ . \\ T \end{array} \begin{bmatrix} n_{i0}^{1} & n_{i1}^{1} \ldots & n_{i,s-1}^{1} \\ n_{i0}^{2} & n_{i1}^{2} \ \ldots & n_{i,s-1}^{2} \\ . & & \\ . & & \\ .n_{i0}^{T} & n_{i1}^{T} \ \ldots & n_{i,s-1}^{T} \end{bmatrix}, \quad i = 0,1,2,\ldots,(s\text{-}1). \tag{2.34}$$

Hence there are T tables with r x r count data. The above matrix shows the data only for the i-th state at time t. By definition of the one-step transition probabilities, the transition that occur at time t, depends on the total counts at time t-1, hence $n_{i.}^{t-1} = \sum_{j=0}^{s-1} n_{ij}^{t}$.

Corresponding to (2.34) the transition probability matrix for time t is

$$\begin{bmatrix} \pi_{00}^{t} & \pi_{01}^{t} \ \ldots & \pi_{0,s-1}^{t} \\ \pi_{10}^{t} & \pi_{11}^{t} \ \ldots & \pi_{1,s-1}^{t} \\ . & & \\ . & & \\ .\pi_{s-1,0}^{t} & \pi_{s-1,1}^{t} \ \ldots & \pi_{s-1,s-1}^{t} \end{bmatrix}, \quad t=1,2,\ldots,T. \tag{2.35}$$

From (2.34) and (2.35), the likelihood function is

$$L = \prod_{t=1}^{T} \prod_{i=0}^{s-1} \frac{n_{i}^{t-1}!}{n_{i0}^{t}!\ldots n_{i,s-1}^{t}!} (\pi_{i0}^{t})^{n_{i0}^{t}} (\pi_{i1}^{t})^{n_{i1}^{t}} \ldots (\pi_{i,s-1}^{t})^{n_{i,s-1}^{t}}. \tag{2.36}$$

From (2.36) the maximum likelihood estimates of the transition probabilities are

$$\hat{\pi}_{ij}^{t} = \frac{n_{ij}^{t}}{n_{i.}^{t-1}}, \quad i, j = 0,1,\ldots,s-1. \tag{2.37}$$

Then using (2.37) we can employ the likelihood ratio test (LRT) for testing the null hypotheses $H_0 : \pi_{ij}^{t} = \pi_{ij}$ (t = 1,…T; i,j = 0,…,s-1) as

$$-2\ln\Lambda = 2\sum_{t=1}^{T} \sum_{i=0}^{s-1} \sum_{j=0}^{s-1} n_{ij}^{t} \ln \frac{n_{ij}^{t}}{n_{i.}^{t-1}\pi_{ij}} \tag{2.38}$$

where π_{ij} represents the transition probabilities for the pooled time.

The expression in (2.38) is χ^2 with (T-1)s(s-1) degrees of freedom under H_0.

For stationarity test, we can consider the null hypotheses $H_0 : \pi_{ij}^{t} = \pi_{ij}^{t+1}$ where

π_{ij}^{t} = transition probability from i to j for order 1 at time t and

π_{ij}^{t+1} = transition probability from i to j for for order 1 at time t+1.

Then

$$-2\ln\Lambda = 2\sum_{t=1}^{T-1}\sum_{i=0}^{s-1}\sum_{j=0}^{s-1} n_{ij}^{t} \ln\frac{n_{ij}^{t}}{n_{i.}^{t-1}\pi_{ij}^{t+1}}. \quad (2.39)$$

Under H_0, (2.39) is χ^2 with (T-2)s(s-1) degrees of freedom.

The above can be extended for a Markov chain of order r. The number of cells in the r-th order Markov chain is s^{r+1} and the number of restrictions is s^r and the corresponding degrees of freedom is $s^{r+1} - s^r = s^r(s-1)$. Here we have assumed all the resulting rows have the constraint of row probability equal to 1, or all the marginal row totals are fixed. If we consider T follow-ups, then the degrees of freedom is $(T-r)s^r(s-1)$. In other words, we have (T-r) sets of tables with s^{r+1} cell frequencies arising from order r Markov model with s states for T follow-ups. The expression of Chi squares remains same as shown below,

$$-2\ln\Lambda = 2\sum_{t=1}^{T-r}\sum_{i=0}^{s^r-1}\sum_{j=0}^{s-1} n_{ij}^{t} \ln\frac{n_{ij}^{t}}{n_{i.}^{t-1}\pi_{ij}^{t+1}}$$

where i denotes the transition types based on the previous transitions.

Hoel (1954) showed that the corresponding degrees of freedom is $s^{r-1}(s-1)^2$ which is not applicable for higher order Markov models because there is no assumption of marginal column totals under Markovian assumptions unless we have a doubly stochastic Markov model, which is only a very special case under restrictive assumptions.

2.9 EXAMPLES

Let us consider some examples employing the data on the Health and Retirement Survey (HRS) conducted during 1992-2004 at two year intervals as mentioned in 1.2.1.1 D.1. In all the waves, mobility index for the elderly population was constructed on the basis of five tasks including walking several blocks, walking one block, walking across the room, climbing several flights of stairs, climbing one flight of stairs. To demonstrate a two-state Markov chain model, we have considered 0 for no difficulty and 1 for difficulty. Table 2.1 displays the pooled number of transitions during two consecutive follow-ups in the period 1992-2004.

Table 2.1. Pooled Transition Counts in Mobility Index (Two States) among Elderly during 1992-2004

	Transition Count			Transition Probability	
States	0	1	$n_{i.}$	0	1
0	22461	5621	28082	0.80	0.20
1	3733	12636	16369	0.23	0.77

Table 2.1 shows all the transitions of first order during the 1992-2004 period. The corresponding transition probabilities are also computed. It is evident that 80 percent of the elderly remained in the state of no difficulty and 77 percent remained in the state of difficulty in the first order transition. The transition probability from no difficulty to difficulty is 0.20 while the transition from some difficulty to no difficulty is 0.23 during the whole period.

We can employ two test statistics for testing the null hypotheses $H_0 : \pi_{ij} = \pi_{ij}^0$, one being the χ^2 test for goodness of fit under null hypothesis

$$\chi^2 = \sum_{i=0}^{s-1} \sum_{j=0}^{s-1} \frac{n_{i.}(\hat{\pi}_{ij} - \pi_{ij}^0)^2}{\pi_{ij}^0}$$

This is χ^2 with s(s-1)-d degrees of freedom under H_0 where d is the number of zero π_{ij}^0's. Here, we may consider the pooled estimates for transition probabilities for testing the null hypothesis.

Similarly, we can also use the likelihood ratio as shown below:

$$-2\ln\Lambda = 2\sum_{i=0}^{s-1} \sum_{j=0}^{s-1} n_{ij} \ln \frac{n_{ij}}{n_{i.}\pi_{ij}^0}.$$

This is χ^2 with s(s-1) degrees of freedom under H_0. This test can also be performed using the pooled estimate of the transition probabilities.

For s-state Markov chain models of order r, we have used the following tests as discussed in section 2.6:

$$\chi^2 = \sum_{i=0}^{s^r-1} \sum_{j=0}^{s-1} \frac{n_{i.}(\hat{\pi}_{ij} - \pi_{ij}^0)^2}{\pi_{ij}^0},$$

and

$$-2\ln\Lambda = 2\sum_{i=0}^{s^r-1} \sum_{j=0}^{s} n_{ij} \ln \frac{n_{ij}}{n_{i.}\pi_{ij}^0},$$

which is χ^2 with $s^r(s-1)$ degrees of freedom under H_0.

Table 2.2 shows the chi-square and likelihood ratio test (LRT) statistics for testing the null hypotheses that all the transition probabilities are 0.5 for first, second, third and fourth order transition probabilities. The results show significant deviation between observed and expected cell counts for all the orders implying rejection of null hypotheses for all the orders.

To test for stationarity, the maximum likelihood estimates of the transition probabilities are:

$$\hat{\pi}_{ij}^t = \frac{n_{ij}^t}{n_{i.}^{t-1}}, \quad i, j = 0,1,\ldots,s-1.$$

Then we can employ the likelihood ratio test for testing the null hypotheses $H_0 : \pi_{ij}^t = \pi_{ij}$ (t=1, … ,T; i,j=0, … ,s-1) as follows

$$-2\ln\Lambda = 2\sum_{t=1}^{T}\sum_{i=0}^{s-1}\sum_{j=0}^{s-1} n_{ij}^t \ln\frac{n_{ij}^t}{n_{i.}^{t-1}\pi_{ij}}.$$

Table 2.2. Test for Inference of Transition Probability for Mobility Index (Two States)

Test-statistics	Value	df	p-value
First Order			
Chi-square	14940.7708	2	0.00000
LRT	15927.3597	2	0.00000
Second Order			
Chi-square	14050.6181	4	0.00000
LRT	15536.7059	4	0.00000
Third Order			
Chi-square	11121.4027	6	0.00000
LRT	12515.5306	6	0.00000
Fourth Order			
Chi-square	7869.84366	8	0.00000
LRT	8953.75616	8	0.00000

This is χ^2 with (T-1)s(s-1) degrees of freedom under H_0. For testing this null hypothesis, we may use the pooled transition probabilities in order to test for transition probabilities.

Alternatively, as mentioned in the previous section, we may consider the null hypotheses $H_0 : \pi_{ij}^t = \pi_{ij}^{t+1}$ (t=1, … ,T; i,j=0, … ,s-1) as follows

$$-2\ln\Lambda = 2\sum_{t=1}^{T-1}\sum_{i=0}^{s-1}\sum_{j=0}^{s-1} n_{ij}^t \ln\frac{n_{ij}^t}{n_{i.}^{t-1}\pi_{ij}^{t+1}}.$$

This is χ^2 with (T-2)s(s-1) degrees of freedom under H_0.

The transition count and transition probability matrices are shown in Table 2.3. Transition probabilities in consecutive waves demonstrate that the transition probabilities remain almost similar in consecutive follow-ups during 1992-2004 for both 0-1 and 1-0. There is no much variation from that of the pooled transition probabilities.

Table 2.3. First Order Transition Count and Probability Matrix for Mobility Index (Two States)

Transition Count Matrix

Follow-up (T)

1	2			3			4			5		
States	0	1	$n_{i.}^1$	0	1	$n_{i.}^2$	0	1	$n_{i.}^3$	0	1	$n_{i.}^4$
0	5239	1389	6628	4340	988	5328	3902	832	4734	3420	826	4246
1	645	1947	2592	703	2196	2899	678	2200	2878	632	2093	2725

Transition Count Matrix

Follow-up (T)

1	6			7		
	0	1	$n_{i.}^5$	0	1	$n_{i.}^6$
0	2947	832	3779	2613	754	3367
1	542	2050	2592	533	2150	2683

Transition Probability Matrix

Follow-up (T)

1	2		3		4		5		6		7	
States	0	1	0	1	0	1	0	1	0	1	0	1
0	0.79	0.21	0.81	0.19	0.82	0.18	0.81	0.19	0.78	0.22	0.78	0.22
1	0.25	0.75	0.24	0.76	0.24	0.76	0.23	0.77	0.21	0.79	0.20	0.80

Table 2.4 indicates that the chi-square for 7 follow-ups is (25.31+4.17+11.15+25.09 +2.07) = 67.79 which is significant (p-value<0.001) with degrees of freedom = (7-1)2(2-1)=12 indicating that the Markov transition probabilities are not stationary.

Table 2.4. Results for Stationarity Test for First Order Mobility Index among Elderly (Two States)					
		Comparison Between Consecutive time points		Comparison with pooled transition matrix	
T	df	Chi-square	p-value	Chi-square	p-value
2	2	25.315	0.000003	9.852	0.007257
3	2	4.167	0.124496	10.744	0.004645
4	2	11.152	0.003788	19.119	0.000071
5	2	25.092	0.000004	1.0773	0.583533
6	2	2.065	0.356091	14.613	0.000671
7	2			25.154	0.000003
Overall χ^2	10,12	67.789	0.000000	80.558	0.000000

The pooled transition counts and transition probabilities are shown in Table 2.5 for second order Markov model. The transition probabilities of the types 0-0-1, 0-1-1, 1-0-1 and 1-1-1 are 0.15, 0.59, 0.48, and 0.87 respectively. It shows that the difficulty in in mobility in two previous follow-ups result in the highest probability to remain in the same state (0.87), followed by difficulty in just prior follow-up (0.59) and difficulty in one follow-up before the prior follow-up (0.48).

Table 2.5. Pooled Transition Counts and Transition Probabilities for Second Order for Mobility Index (Two States)						
		Transition Count			Transition Probability	
		0	1	$n_{i.}$	0	1
0	0	15687	2833	18520	0.85	0.15
0	1	1825	2589	4414	0.41	0.59
1	0	1535	1399	2934	0.52	0.48
1	1	1263	8098	9361	0.14	0.87

The test for the first order can be extended for a Markov chain of order r. The number of cells in the r-order Markov chain is s^{r+1} and the number of restrictions is s^r and the corresponding degrees of freedom is $s^{r+1} - s^r = s^r(s-1)$. Here we have assumed all the resulting rows have the constraint of row probability equal to 1, or all the marginal row totals are fixed. If we consider T follow-ups, then the degrees of freedom is $(T-r)s^r(s-1)$ for null hypothesis based on pooled transitions. In other words, we have (T-r) sets of tables with s^{r+1} cell frequencies arising from order r Markov model with s states for T follow-ups as shown below:

$$-2\ln\Lambda = 2\sum_{t=1}^{T-r}\sum_{i=0}^{s^r-1}\sum_{j=0}^{s-1} n_{ij}^{t}\ln\frac{n_{ij}^{t}}{n_{i.}^{t-1}\pi_{ij}^{t+1}}$$

Similarly, the degrees of freedom is $(T-r-1)s^r(s-1)$ for the test based on subsequent transition probability due to non availability of transition probability for (T+1) and the test statistic is:

$$-2\ln\Lambda = 2\sum_{t=1}^{T-r-1}\sum_{i=0}^{s^r-1}\sum_{j=0}^{s-1} n_{ij}^{t}\ln\frac{n_{ij}^{t}}{n_{i.}^{t-1}\pi_{ij}^{t+1}}.$$

Table 2.6. Second Order Transition Count and Probability Matrix for Mobility Index (Two States)

Transition Count

Follow-up (T)

1	2	3			4			5			6		
States		0	1	$n_{i.}^{1}$	0	1	$n_{i.}^{2}$	0	1	$n_{i.}^{3}$	0	1	$n_{i.}^{4}$
0	0	4209	761	4970	3549	544	4093	3045	522	3567	2619	541	3160
0	1	493	790	1283	388	526	914	326	415	741	322	418	740
1	0	311	285	596	371	283	654	322	287	609	292	293	585
1	1	247	1534	1781	289	1719	2008	296	1644	1940	216	1629	1845

Transition Count

Follow-up (T)

1	2	7		
		0	1	$n_{i.}^{5}$
0	0	2265	465	2730
0	1	296	440	736
1	0	239	251	490
1	1	215	1572	1787

Transition Probability Matrix

Follow-up (T)

1	2	3		4		5		6		7	
States		0	1	0	1	0	1	0	1	0	1
0	0	0.85	0.15	0.87	0.13	0.85	0.15	0.83	0.17	0.83	0.17
0	1	0.38	0.62	0.42	0.58	0.44	0.56	0.44	0.56	0.40	0.60
1	0	0.52	0.48	0.57	0.43	0.53	0.47	0.50	0.50	0.49	0.51
1	1	0.14	0.86	0.14	0.86	0.15	0.85	0.12	0.88	0.12	0.88

The stationarity test results are shown in Tables 2.6 and 2.7. The likelihood ratio test based on the null hypothesis using the pooled estimates shows that the Chi square value is 61.608 and the corresponding degrees of freedom is $(T-r)s^r(s-1)$ =(7-2)(4)(2-1)=20.

The second order transition probabilities do not show stationarity over time on the basis of this test. The test based on subsequent transition probabilities also show similar results (chi square= 87.003 with 16 degrees of freedom).

Table 2.7. Results for Stationarity Test for Second Order Mobility Index among Elderly (Two States)

		Comparison between Consecutive Time Points		Comparison with Pooled Transition Matrix	
T	df	Chi-square	p-value	Chi-square	p-value
3	4	30.886303	0.000003	4.765634	0.312201
4	4	12.050641	0.016979	20.148281	0.000467
5	4	40.244060	0.000000	8.433673	0.076923
6	4	3.821964	0.430636	15.875612	0.003191
7	4			12.385002	0.014707
Overall χ^2	16, 20	87.002968	0.000000	61.608202	0.000004

Table 2.8 shows that the chi-square value for testing the stationarity of the third and fourth order transition probabilities based on the subsequent transition probabilities. The chi-square values are (29.93+42.15+13.72=85.8) and (57.30+36.35=93.65) respectively for third and fourth orders. The degrees of freedom for the third and the fourth orders are (7-3-1)(8)(2-1)=24 and (7-4-1)(16)(2-1)=32 respectively. The p-values for the test based on subsequent transition probabilities indicate that transition probabilities of third and fourth orders are also significant. However, fourth order shows nonsignificant result accepting the hypothesis of stationarity if the null hypothesis values are pooled transitions over time.

Table 2.8. Results for Stationary Test for Third and Fourth Orders (Two States)

		Comparison between Consecutive Time Points		Comparison with Pooled Transition Matrix	
T	df	Chi-square	p-value	Chi-square	p-value
	Third Order				
4	8	29.932	0.000217	21.614	0.005685
5	8	42.156	0.000001	11.603	0.169838
6	8	13.725	0.089221	11.973	0.152408
	8			16.098	0.040996

Table 2.8. (Continued)

		Comparison between Consecutive Time Points		Comparison with Pooled Transition Matrix	
T	df	Chi-square	p-value	Chi-square	p-value
Overall χ^2	24, 32	85.813	0.000000	61.287	0.001383
	Fourth Order				
5	16	57.302	0.000001	17.907	0.329359
6	16	36.354	0.002583	13.422	0.641687
7	16			13.521	0.634352
Overall χ^2	32,48	93.656	0.000000	44.850	0.602684

Now let us consider mobility index for defining three states: state 1= no difficulty= mobility index=0, state 2= little difficulty= mobility index = 1, state 3= more difficulty = mobility index = 2+. The pooled transition counts and transition probabilities are shown in Table 2,9. It is evident from Table 2.9 that the probability of remaining in states 1, 2 and 3 in consecutive follow-ups are 0.80, 0.39 and 0.71 respectively. In other words, those with little difficulty move either to no difficulty or more difficulty at a higher rate.

2.9. Pooled Transition Matrix for First Order Markov Chain (Three States)

	Transition Count				Transition Probability		
States	0	1	2+	$n_{i.}$	0	1	2+
0	22461	3824	1797	28082	0.80	0.14	0.06
1	2789	3050	2018	7857	0.36	0.39	0.26
2+	944	1496	6072	8512	0.11	0.18	0.71

The test results for the first, second and third orders are shown in Table 2.10 for testing the null hypothesis $H_0 : \pi_{ij} = \pi_{ij}^0$ .Then for the Chi square and likelihood ratio tests we can use expressions 2.79 and 2.80 respectively. The results are displayed in Table 2.10 for null hypothesis value of 1/3. It appears that the null hypothesis can be rejected for the first, second and third order models.

Example 2.10. Test for Inference of Transition Probability for Mobility Index (Three States)

Test-statistics	**Value**	**df**	**p-value**
First Order			
Chi-square	33525.5476	6	0.00000
LRT	32015.0622	6	0.00000
Second Order			

Example 2.10. (Continued)			
Test-statistics	**Value**	**df**	**p-value**
Chi-square	29115.8401	12	0.00000
LRT	28268.3987	12	0.00000
Third Order			
Chi-square	22575.3217	18	0.00000
LRT	22081.6341	18	0.00000

Table 2.11. Results for Stationary Test for Three States Mobility Index for Third and Fourth Order (Three States)

		Comparison between Consecutive Time Points		Comparison with Pooled Transition Matrix	
T	df	Chi-square	p-value	Chi-square	p-value
	First Order				
2	6	32.784013	0.000012	19.662893	0.003179
3	6	22.694355	0.000906	13.583227	0.034655
4	6	16.794682	0.010068	24.323891	0.000455
5	6	23.560164	0.000629	3.269261	0.774372
6	6	5.575859	0.472336	14.470554	0.024799
7	6			25.532697	0.000272
Overall χ^2	30, 36	101.409073	0.000000	100.842523	0.000000
	Second Order				
3	18	84.814384	0.000000	29.901295	0.038417
4	18	45.207600	0.000387	31.354179	0.026186
5	18	65.707112	0.000000	23.615138	0.168045
6	18	25.173109	0.120222	22.833023	0.197088
7	18			20.159860	0.323892
Overall χ^2	72, 90	220.902206	0.000000	127.863496	0.005385
	Third Order				
4	54	121.539439	0.000000	49.874225	0.634188
5	54	153.021052	0.000000	48.204504	0.696435
6	54	115.080365	0.000003	47.826211	0.710045
7	54			53.652862	0.487728
Overall χ^2	162,216	389.640856	0.000000	199.557802	0.782245

Table 2.11 shows the likelihood ratio tests for the first, second and third orders for three-state mobility transitions among elderly. The degrees of freedom for the chi-squares can be obtained from $(T-r-1)s^r(s-1)$, ie., (7-1-1)(3)(3-1) for first order, (7-2-1)(9)(3-1) for second order and (7-3-1)(27)(3-1) for third order for the testing of null hypothesis that two consecutive transition probabilities over time have same values. In this case, the transition probabilities are not stationary even at the third order. However, if we consider the pooled transition probabilities, then the third order transition probabilities appear to be stationary.

2.10 SUMMARY

This chapter includes a brief review of some basic concepts of probability, stochastic processes and Markov chain. In addition, some test procedures are illustrated for testing the hypothesis on transition probabilities and stationarity of Markov chains. For test on transition probabilities, the chi-square test of goodness of fit and the likelihood ratio method are illustrated. In testing for stationarity, two different test procedures are employed, one is based on the traditional way of using the pooled counts and the other is based on the consecutive follow-up counts. The maximum likelihood estimates, test procedures, order of Markov chain or test for stationarity can be obtained from several studies (Bartlett, 1951; Hoel, 1954; Good, 1955; Anderson and Goodman, 1957; Goodman, 958a, 1958b, 1958c; Billinsley, 1961; Gold, 1963; Miller, 1963; Regier, 1968; Guthrie and Youssef, 1970; Duncan and Lin, 1972; Chatfield, 1973; Katz, 1981; Kelton and Kelton, 1984; Sundberg, 1986; Reeves, 1993). Bhat (1971) presented the estimation and the test procedures for Markov chains.

Chapter 3

GENERALIZED LINEAR MODELS AND LOGISTIC REGRESSION

3.1 INTRODUCTION

In a generalized linear model (GLM), we can specify linear and nonlinear models through a unified approach. It is also important that the GLM allows modeling of non-normal response distributions. Hence, the application of GLM increases manifold due to its potential use in situations where normality assumption is not satisfied. As a general principle for a GLM, we assume that the response variable follows a distribution which is a member of an exponential family. The exponential family includes normal, Poisson, binomial, exponential, and gamma distributions. It is noteworthy that the linear models based on the assumption of normality of error distributions can be demonstrated as a special case of the GLM. In this chapter, we introduce the GLM as a general approach on the basis of the theoretical exposition demonstrated in the seminal work of McCullagh and Nelder (1989) and we also provide a review of the logit model using the logit link function for binary or polytomous outcome data. In addition, the likelihood method of estimating the parameters of a logistic regression model is also reviewed. These models are reviewed in this chapter in order to provide the necessary background for the models which will be described in the subsequent chapters of this book.

3.2 REVIEW OF THE GENERALIZED LINEAR MODELS

A review of the important features of the generalized linear model is presented in this section. The regression part of the GLM can be shown through link functions depending on the underlying pattern of relationship between the dependent variable and covariates for both normal or non-normal outcomes. A wide range of functions can be included for modeling under GLM. The generalized linear models cover ordinary regression models, logistic regression model, log-linear model and Poisson regression model among a number of potential models.

3.2.1 The Components of a Generalized Linear Model

In the classical linear model, $Y = X\beta + \varepsilon$ where $Y = (Y_1,...,Y_n)'$ is the vector of observations on the response variable, $\beta = (\beta_0,...,\beta_p)'$ is the vector of parameters, $\varepsilon = (\varepsilon_1,...,\varepsilon_n)'$ is the error vector and X is the $n \times (p+1)$ matrix of known constants

$$X = \begin{bmatrix} x_1' \\ x_2' \\ . \\ . \\ . \\ x_n' \end{bmatrix} = \begin{bmatrix} 1 & x_{11} & x_{12} & . & . & . & x_{1p} \\ 1 & x_{21} & x_{22} & . & . & . & x_{2p} \\ . & . & . & . & . & . & . \\ . & . & . & . & . & . & . \\ . & . & . & . & . & . & . \\ 1 & x_{n1} & x_{n2} & . & . & . & x_{np} \end{bmatrix}. \quad (3.1)$$

In regression set-ups the rows x_i' in (3.1) provide the levels of the covariates or independent variables $x' = (1, x_1,..., x_p)$ corresponding to the i-th observation on the response variable (i= 1, .., n). It is assumed that the errors are zero mean uncorrelated random variables with a constant variance, σ^2. Further specification of the model involves the stronger distributional assumption that the errors are also normally distributed.

The generalized linear model, which incorporates the classical model described above as a special case, comprises of three components, namely the systematic component, the random component and a link function connecting the two, as follows (McCullagh and Nelder, 1989):

1. The random component assumes that Y_i 's have independent distributions with $E(Y_i) = \mu_i$ and a constant variance σ^2.
2. The systematic component shows that the vector of values of the covariates $x_i' = x_{i0}(\equiv 1), x_{i1}, x_{i2},...., x_{ip}$ produces a linear predictor η_i given by

$$\eta_i = \sum_{j=0}^{p} x_{ij}\beta_j = x_i'\beta .$$

3. The link function $g(.)$ which shows the relationship between the random and systematic components as $\eta_i = g(\mu_i)$.

For the classical linear model the link function is the identity function $g(\mu) = \mu$.

Let us consider the following distribution of Y:

$$f_Y(y;\theta,\phi) = e^{\{(y\theta - b(\theta))/a(\phi) + c(y,\phi)\}} \quad (3.2)$$

for some specific functions a(.), b(.) and c(.). If ϕ is known, (3.2) is an exponential family model with canonical parameter θ.

The expectation and variance of Y can be obtained as follows:

(i) $E(y) = \dfrac{db(\theta)}{d\theta}$

(ii) $Var(Y) = \dfrac{dE(Y)}{d\theta} = \dfrac{d^2b(\theta)}{d\theta^2}$ depends only on the canonical parameter (and hence on the mean) and will be called the variance function where the variance function is considered as a function of μ and will be written as $V(\mu)$;

(iii) $a(\phi)$ is independent of θ and depends only on ϕ.

The function $a(\phi)$ is commonly of the form $a(\phi) = \phi / w$, where ϕ, also denoted by σ^2 and called the dispersion parameter is constant over observations, and w is a known prior weight that varies from observation to observation.

The link function relates the linear predictor η to the expected value μ of an observation y. Models for counts based on independence in cross-classified data lead naturally to multiplicative effects, and this is expressed by the log-link, $\eta = \log_e \mu$, with its inverse $\mu = e^{\eta}$. Now additive effects contributing to η become multiplicative effects contributing to μ and μ is necessarily positive. The logit link function can be expressed as $\eta = \log_e \mu/(1-\mu)$.

3.2.2 Parameter Estimation: Logit Link Function in the Generalized Linear Model

For any binomial variable, Y, the probability mass function can be expressed as

$$f_Y(y;\theta,\phi) = \binom{n}{y} \pi^y (1-\pi)^{n-y} = e^{\{\ln\binom{n}{y} + y\ln\pi + (n-y)\ln(1-\pi)\}}$$

$$= e^{\{\ln\binom{n}{y} + y\ln\pi + n\ln(1-\pi) - y\ln(1-\pi)\}} = e^{\{y\ln\left[\frac{\pi}{1-\pi}\right] + n\ln(1-\pi) + \ln\binom{n}{y}\}}. \quad (3.3)$$

Therefore, from (3.3), for the binomial distribution

$$\theta = \ln\left[\pi/(1-\pi)\right],\ \pi = e^{\theta}/(1+e^{\theta}),\ b(\theta) = -n\ln(1-\pi),$$

$$a(\phi) = 1,\ c(y,\phi) = \ln\binom{n}{y},\ E(y) = \frac{db(\theta)}{d\theta} = \frac{db(\theta)}{d\pi}.\frac{d\pi}{d\theta},$$

where

$$\frac{d\pi}{d\theta} = \frac{e^{\theta}}{1+e^{\theta}} - \left[\frac{e^{\theta}}{1+e\theta}\right]^2 = \pi(1-\pi). \tag{3.4}$$

Therefore, it follows from (3.4) that

$$E(y) = \left(\frac{n}{1-\pi}\right)\pi(1-\pi) = n\pi. \tag{3.5}$$

We recognize (3.5) as the mean of the binomial distribution. Also, using (3.5),

$$Var(y) = \frac{dE(y)}{d\theta} = \frac{dE(y)}{d\pi}\frac{d\pi}{d\theta} = n\pi(1-\pi), \tag{3.6}$$

and (3.6) gives the variance of a binomial distribution. For the exponential family, the log likelihood function corresponding to a random sample of size n is

$$l(y,\beta) = \sum_{i=1}^{n} [\{y_i\theta_i - b(\theta_i)\}/a(\phi) + c(y_i,\phi)].$$

Thus for the canonical link in the binomial case, we have $\eta_i = g[E(y_i)] = g(\mu_i) = \ln[\mu_i/(1-\mu_i)] = x_i'\beta = \theta_i$ and $\mu_i = \pi_i$ where x_i' is the i-th row of the X-matrix. Therefore,

$$\frac{\partial l}{\partial \beta} = \frac{\partial l}{\partial \theta_i}.\frac{\partial \theta_i}{\partial \beta} = \frac{1}{a(\phi)}\sum_{i=1}^{n}[y_i - \frac{db(\theta_i)}{d\theta_i}]x_i = \frac{1}{a(\phi)}\sum_{i=1}^{n}[y_i - \mu_i]x_i.$$

Consequently, we can find the maximum likelihood estimates of the parameters by solving the system of equations

$$\frac{1}{a(\phi)}\sum_{i=1}^{n}[y_i - \mu_i]x_i = 0. \tag{3.7}$$

For the binomial distribution $a(\varphi) = 1$, so these equations (3.7) become

$$\sum_{i=1}^{n}[y_i - \mu_i]x_i = 0. \tag{3.8}$$

This is actually a system of p +1 equations, one for each model parameter. In matrix form, these equations may be written as

$$X'(y-\mu) = 0 \tag{3.9}$$

where $\mu' = [\mu_1, \mu_2, ..., \mu_n]$. These are called the maximum likelihood score equations.

To solve the score equations (3.8) or (3.9), we can use iteratively reweighted least squares (IRLS) algorithm. We start by finding a first-order Taylor series approximation in the neighborhood of the solution

$$y_i - \mu_i \approx \frac{d\mu_i}{d\eta_i}(\eta_i^* - \eta_i) \ (i = 1,2,...,n).$$

Now for a canonical link $\eta_i = \theta_i$, and

$$y_i - \mu_i \approx \frac{d\mu_i}{d\theta_i}(\eta_i^* - \eta_i)(i = 1,2,...,n). \tag{3.10}$$

Here, η_i^* is not known and can be replaced by the working dependent variable, z_i, and thus, we can express

$$z_i = \hat{\eta}_i + (y_i - \hat{\mu}_i)\frac{d\eta_i}{d\mu_i}.$$

From (3.10), we obtain

$$z_i - \hat{\eta}_i \approx (y_i - \mu_i)\frac{d\theta_i}{d\mu_i}$$

and the variance of $\hat{\eta}_i$

$$Var(\hat{\eta}_i) = \left(\frac{d\theta_i}{d\mu_i}\right)^2 Var(y_i)$$

where $\frac{d\theta_i}{d\mu_i} = \frac{1}{\text{var}(\hat{\mu}_i)}$ and $Var(y_i) = \text{var}(\hat{\mu}_i)a(\phi)$.

If we let V be a n x n diagonal matrix whose diagonal elements are the $\text{var}(\hat{\eta}_i)$ then using matrix notation (3.10) can be written as

$$y - \mu = V^{-1}(\eta^* - \eta).$$

We may then rewrite the score equations for $\eta_i = g[E(y_i)] = g(\mu_i) = \ln[\mu_i/(1-\mu_i)] = x_i'\beta = \theta_i \quad (i = 1,2,...,n)$ as follows:

$$X'(y - \mu) = 0,\ X'V^{-1}(\eta^* - \eta) = 0,\ X'V^{-1}(\eta^* - X\beta) = 0.$$

Thus, the maximum likelihood estimate of β is

$$\hat{\beta} = (X'V^{-1}X)^{-1}X'V^{-1}z. \tag{3.11}$$

It is interesting to note similarity of (3.11) to the expression obtained in standard regression model.

3.3 Likelihood Estimation of Logistic Regression Models

The logistic regression is shown in the previous section as a link function in the generalized linear model. In this section, the general procedure for logistic regression models is discussed.

Let Y =1, if an event occurs during a defined study period,
=0, otherwise.

Many distribution functions have been proposed for use in the analysis of a dichotomous outcome variable. The logistic regression is a natural choice for many survival or reliability problems due to its functional form as well as for the underlying utility of meaningful interpretation relating covariates and outcome variable of interest.

Let us consider a simple logistic regression first involving the outcome variable, Y, and a single covariate, X. We can define the probability of Y=1 for given X and probability of Y=0 for given X as shown below:

$$\pi(X) = P(Y=1 \mid X) = e^{\beta_0+\beta_1 X} / (1+e^{\beta_0+\beta_1 X}) \quad (3.12)$$

and

$$1-\pi(X) = P(Y=0 \mid X) = 1 - e^{\beta_0+\beta_1 X} / (1+e^{\beta_0+\beta_1 X}) = 1/(1+e^{\beta_0+\beta_1 X}). \quad (3.13)$$

A transformation of $\pi(X)$ that is central to our study of logistic regression is the logit transformation. This transformation is defined, in terms of $\pi(X)$, as $g(X) = \ln[\pi(X)/\{1-\pi(X)\}]$. Then using (3.12) and (3.13) we get

$$g(X) = \ln\left[\frac{\pi(X)}{1-\pi(X)}\right] = \ln\left[\frac{e^{\beta_0+\beta_1 X} / (1+e^{\beta_0+\beta_1 X})}{1/(1+e^{\beta_0+\beta_1 X})}\right] = \ln\left[e^{\beta_0+\beta_1 X}\right] = \beta_0 + \beta_1 X. \quad (3.14)$$

The importance of this transformation is that g(X), as seen in (3.14), has many of the desirable properties of a linear regression model. The logit, g(X), is linear in its parameters, may be continuous, and may range from $-\infty$ to ∞, depending on the range of X.

3.3.1 Fitting of Logistic Regression Model

Suppose we have a sample of n independent observations of the pair (X_i, Y_i), 1=1,2,…,n, where Y_i denotes the value of a dichotomous outcome variable and X_i is the value of the independent variable for the i-th subject. Furthermore, assume that the outcome variable has been coded as 0 or 1, representing absence or presence of characteristic, respectively. Fitting the logistic regression model to a set of data requires that we estimate the values of β_0 and β_1, the unknown parameters.

The contribution to the likelihood function for the pair (X_i, Y_i) is through the term

$$L_i(X_i) = [\pi(X_i)]^{Y_i} [1-\pi(X_i)]^{1-Y_i}. \quad (3.15)$$

Since the observations are assumed to be independent, the likelihood function is the product of the terms given by (3.15) for $i = 1, 2, ..., n$ and can be expressed as

$$L(\beta) = \prod_{i=1}^{n} L_i(X_i) = \prod_{i=1}^{n} [\pi(X_i)]^{Y_i} [1-\pi(X_i)]^{1-Y_i}$$

$$= \prod_{i=1}^{n} \left[\frac{e^{\beta_0+\beta_1 X_i}}{1+e^{\beta_0+\beta_1 X_i}}\right]^{Y_i} \left[\frac{1}{1+e^{\beta_0+\beta_1 X_i}}\right]^{1-Y_i}. \quad (3.16)$$

From (3.16), the log likelihood function is

$$\ln L(\beta) = \sum_{i=1}^{n} \left[Y_i \ln\left\{\frac{e^{\beta_0+\beta_1 X_i}}{1+e^{\beta_0+\beta_1 X_i}}\right\} + (1-Y_i)\ln\left\{\frac{1}{1+e^{\beta_0+\beta_1 X_i}}\right\}\right]$$

$$= \sum_{i=1}^{n} \left[Y_i \left\{ (\beta_0 + \beta_1 X_i) - \ln(1 + e^{\beta_0 + \beta_1 X_i}) \right\} - (1 - Y_i) \ln(1 + e^{\beta_0 + \beta_1 X_i}) \right]$$

$$= \sum_{i=1}^{n} \left[Y_i (\beta_0 + \beta_1 X_i) - \ln(1 + e^{\beta_0 + \beta_1 X_i}) \right]. \tag{3.17}$$

Differentiating (3.17) with respect to β_0 and setting the derivative equal to zero, we get

$$\frac{\partial \ln L(\beta)}{\partial \beta_0} = \sum_{i=1}^{n} \left[Y_i - \frac{e^{\beta_0 + \beta_1 X_i}}{1 + e^{\beta_0 + \beta_1 X_i}} \right] = 0 \text{ or } \sum_{i=1}^{n} [Y_i - \pi(X_i)] = 0. \tag{3.18}$$

Similarly differentiating (3.17) with respect to β_1 and setting the derivative equal to zero gives

$$\frac{\partial \ln L(\beta)}{\partial \beta_1} = \sum_{i=1}^{n} \left[X_i Y_i - \frac{X_i e^{\beta_0 + \beta_1 X_i}}{1 + e^{\beta_0 + \beta_1 X_i}} \right] = 0,$$

$$\text{or } \frac{\partial \ln L(\beta)}{\partial \beta_1} = \sum_{i=1}^{n} X_i \left[Y_i - \frac{e^{\beta_0 + \beta_1 X_i}}{1 + e^{\beta_0 + \beta_1 X_i}} \right] = 0,$$

$$\text{or } \sum_{i=1}^{n} X_i [Y_i - \pi(X_i)] = 0. \tag{3.19}$$

Solving the equations (3.18) and (3.19) we obtain the estimates for β_0 and β_1.

3.3.2 Tests

To test the hypothesis that $H_0 : \beta_1 = 0$ a likelihood ratio test can be used. The test statistic

$$\chi^2 = -2[\ln L(\hat{\beta}_0) - \ln L(\hat{\beta}_0, \hat{\beta}_1)] \tag{3.20}$$

is distributed as chi-square with 1 degree of freedom under the hypothesis that β_1 is equal to zero.

An alternative test for the significance of the coefficients is the Wald test using the test statistic:

$$W = \hat{\beta}_1 / se(\hat{\beta}_1) \tag{3.21}$$

which follows the standard normal distribution under the null hypothesis $H_0 : \beta_1 = 0$. Though the Wald test, given by (3.21), is used by many, it is less powerful than the likelihood ratio test given by (3.20). The Wald test often misleads the user to conclude that the coefficient (consequently the corresponding risk factor) is not significant when it indeed is.

3.4 The Multiple Logistic Regression Model

Let X be a vector of p independent variables $X = \left[X_1, X_2, ..., X_p\right]'$. Then

$$\pi(X) = e^{g(X)} / (1 + e^{g(X)}) \tag{3.22}$$

where $g(X) = \beta_0 + \beta_1 X_1 + ... + \beta_p X_p$.

Equation (3.22) generalizes the logit given by (3.14). If some of the independent variables are nominal scaled variables such as race, sex, treatment group, and so fourth, then it is inappropriate to include them in the model as if they were interval scaled. In this situation the method of choice is to use a collection of design variables (or dummy variables).

3.4.1 Fitting the Multiple Logistic Regression Model

Assume that we have a sample of n independent observations of the pair (X_i, Y_i) where

$X_i = \left[X_{i1}, X_{i2}, ..., X_{ip}\right]'$ = observations on the p independent variables for the i-th individual. Hence the fitting of the model requires estimating the parameter vector

$$\beta = \left[\beta_0, \beta_1, ..., \beta_p\right]'.$$

The likelihood function is identical to that given for a single independent variable, with the only change being that $\pi(X)$ is defined in terms of a multivariate logit. There will be (p+1) likelihood equations which are obtained by differentiating the log likelihood function with respect to the (p+1) coefficients. The likelihood equation that result may be expressed as

$$\sum_{i=1}^{n} [Y_i - \pi(X_i)] = 0 \text{ and } \sum_{i=1}^{n} X_{ij}[Y_i - \pi(X_i)] = 0, \text{ j=1,2,.....,p.} \tag{3.23}$$

Solving (3.23) we obtain the maximum likelihood estimates of the regression parameters. The variances and covariances of the estimated coefficients can be obtained from

$$I^*_{ju} = -\sum_{i=1}^{n} X_{ij} X_{iu} \pi_i (1 - \pi_i) \tag{3.24}$$

for j,u=0,1,...,p where $\pi_i = \pi(X_i)$. Let $I_{ju} = (-1) I^*_{ju}$ with I^*_{ju} given by (3.24). Then the information matrix is defined by $I(\beta)$ where (j,u)-th element of $I(\beta)$ is I_{ju}.

The variances and covariances of the estimated coefficients are obtained from the inverse of the information matrix, i.e.

$$\Sigma(\beta) = I^{-1}(\beta) \tag{3.25}$$

where $\hat{\sigma}^2(\hat{\beta}_j)$ is the j-th diagonal element of $\hat{\Sigma}(\beta)$ and $\hat{\sigma}(\hat{\beta}_j, \hat{\beta}_u)$ is the (j,u)-th element of $\hat{\Sigma}(\beta)$. $\hat{\sigma}^2(\hat{\beta}_j)$ is the estimated variance of $\hat{\beta}_j$ and $\hat{\sigma}(\hat{\beta}_j, \hat{\beta}_u)$ is the estimated covariance of $\hat{\beta}_j$ and $\hat{\beta}_u$, and

$$se(\hat{\beta}_j) = [\hat{\sigma}^2(\hat{\beta}_j)]^{1/2}.$$

The information matrix used in (3.25) can be expressed as

$$\hat{I}(\hat{\beta}) = X'VX \qquad (3.26)$$

where X is an n x (p+1) matrix containing the data for all subjects, i.e.,

$$X = \begin{bmatrix} 1 \;\; X_{11} \;\; \ldots\ldots\ldots X_{1p} \\ 1 \;\; X_{21} \;\; \ldots\ldots\ldots\ldots X_{2p} \\ \ldots\ldots\ldots\ldots\ldots\ldots\ldots\ldots \\ \ldots\ldots\ldots\ldots\ldots\ldots\ldots\ldots \\ 1 \;\; X_{n1} \;\; \ldots\ldots\ldots\ldots X_{np} \end{bmatrix} \quad \text{and} \quad V = \begin{bmatrix} \hat{\pi}_1(1-\hat{\pi}_1),\ 0, \ldots\ldots\ldots\ldots,\ 0 \\ 0,\ \hat{\pi}_2(1-\hat{\pi}_2),\ 0,\ \ldots\ldots,\ 0 \\ \ldots\ldots\ldots\ldots\ldots\ldots\ldots\ldots\ldots\ldots \\ \ldots\ldots\ldots\ldots\ldots\ldots\ldots\ldots\ldots\ldots \\ 0, \ldots\ldots \ldots\ldots,\ 0, \hat{\pi}_n(1-\pi_n) \end{bmatrix}.$$

3.4.2 Testing for the Significance of the Model

An approximate 100(1-α) percent confidence interval for β_j can be obtained as

$$\hat{\beta}_j \pm z_{\alpha/2}\sqrt{I_{jj}^{-1}}$$

where $z_{\alpha/2}$ is the 100(1-α /2) percentile of the standard normal distribution.

To test the hypothesis that some of the β_j's are zero, a likelihood ratio test can be used. For testing the null hypothesis

$$H_0 : \beta_1 = \beta_2 = \ldots\ldots\ldots = \beta_p = 0$$

we can use the test statistic

$$\chi^2 = -2[\ln L(\hat{\beta}_0) - \ln L(\hat{\beta}_0, \hat{\beta}_1, \ldots\ldots, \hat{\beta}_p)]. \qquad (3.27)$$

Under H_0, (3.27) has a χ_p^2 distribution.

3.4.3 Interpreting Coefficients as Odds Ratios

Let us consider the case where there are three independent variables X_1, X_2, X_3, and

$\hat{\pi}(X)$ = estimated probability of the occurrence of an event

$$= \frac{e^{\hat{\beta}_0+\hat{\beta}_1 X_1+\hat{\beta}_2 X_2+\hat{\beta}_3 X_3}}{1+e^{\hat{\beta}_0+\hat{\beta}_1 X_1+\hat{\beta}_2 X_2+\hat{\beta}_3 X_3}}$$

and

$1-\hat{\pi}(X)$ = estimated probability of the non-occurrence of the event

$$= \frac{1}{1+e^{\hat{\beta}_0+\hat{\beta}_1 X_1+\hat{\beta}_2 X_2+\hat{\beta}_3 X_3}}.$$

Then estimated odds of occurrence of the event for given X is

$$\frac{\hat{\pi}(X)}{1-\hat{\pi}(X)} = \frac{\dfrac{e^{\hat{\beta}_0+\hat{\beta}_1 X_1+\hat{\beta}_2 X_2+\hat{\beta}_3 X_3}}{1+e^{\hat{\beta}_0+\hat{\beta}_1 X_1+\hat{\beta}_2 X_2+\hat{\beta}_3 X_3}}}{\dfrac{1}{1+e^{\hat{\beta}_0+\hat{\beta}_1 X_1+\hat{\beta}_2 X_2+\hat{\beta}_3 X_3}}} = e^{\hat{\beta}_0+\hat{\beta}_1 X_1+\hat{\beta}_2 X_2+\hat{\beta}_3 X_3}.$$

The estimated relative odds of occurrence of an event comparing those for whom X_1 is present (X_1=1) with those for whom X_1 is absent (X_1=0) is

$$\text{Odds Ratio } (OR) = \frac{e^{\hat{\beta}_0+\hat{\beta}_1 .1+\hat{\beta}_2 X_2+\hat{\beta}_3 X_3}}{e^{\hat{\beta}_0+\hat{\beta}_1 .0+\hat{\beta}_2 X_2+\hat{\beta}_3 X_3}} = e^{\hat{\beta}_1} \text{ and } \ln OR = \hat{\beta}_1.$$

In this example, the levels of X_2 and X_3 are same for both $X_1 = 0$ and $X_1 = 1$.

3.4.4 Polytomous Logistic Regression

Let us suppose that the dependent variable Y may take on nominal scale values 0, 1, 2,…,K-1. The logistic regression model for a binary outcome variable was parameterized in terms of the logit of Y=1 versus Y=0. In the K category model we have (K-1) logit functions:

(i) Y=1 versus Y=0;
(ii) Y=2 versus Y=0
…
(K-1) Y=K-1 versus Y=0.

The group coded Y=0 will serve as the reference outcome value. The logit for comparing Y=k to Y=k-1 may be obtained as the difference between the logit of Y=k versus Y=0 and the logit of Y=k-1 versus Y=0.

Let X be the vector of covariates $X_1, X_2, \ldots, X_p$ with X_0=1 to account for the constant term. Then let $g_0(X) = 0$ and

$$g_1(X)=\ln\left[\frac{P(Y=1|X)}{P(Y=0|X)}\right]=\beta_{10}+\beta_{11}X_1+.......+\beta_{1p}X_p=X'\beta_1.$$

Similarly,

$$g_2(X)=\ln\left[\frac{P(Y=2|X)}{P(Y=0|X)}\right]=\beta_{20}+\beta_{21}X_1+.......+\beta_{2p}X_p=X'\beta_2$$

and, in general,

$$g_k(X)=\ln\left[\frac{P(Y=k|X)}{P(Y=0|X)}\right]=\beta_{k0}+\beta_{k1}X_1+.......+\beta_{kp}X_p=X'\beta_k,$$

k=1,2,…,K-1.

The conditional probabilities are

$$P(Y=0|X)=\frac{1}{1+e^{g_1(X)}+e^{g_2(X)}+...+e^{g_k(X)}+...+e^{g_{K-1}(X)}}=\pi_0(X),$$

$$P(Y=1|X)=\frac{e^{g_1(X)}}{1+e^{g_1(X)}+e^{g_2(X)}+...+e^{g_k(X)}+...+e^{g_{K-1}(X)}}=\pi_1(X),$$

$$P(Y=2|X)=\frac{e^{g_2(X)}}{1+e^{g_1(X)}+e^{g_2(X)}+...+e^{g_k(X)}+...+e^{g_{K-1}(X)}}=\pi_2(X)$$

and, in general terms,

$$P(Y=k|X)=\frac{e^{g_k(X)}}{1+e^{g_1(X)}+e^{g_2(X)}+...+e^{g_k(X)}+...+e^{g_{K-1}(X)}}=\pi_k(X)$$

k= 0,1,2,…,K-1.

Each of these is a function of the vector of (K-1)(p+1) parameters

$$\beta=\begin{pmatrix}\beta_1\\ \beta_2\\ .\\ .\\ .\\ \beta_{K-1}\end{pmatrix}.$$

The general expression is

$$P(Y=j|X)=\frac{e^{g_j(X)}}{\sum_{k=0}^{K-1} e^{g_k(X)}}, \qquad j=1,2,...,K-1 \qquad (3.28)$$

where $\beta_0=0$ and $g_0(X)=0$. Note that (3.28) is a generalization of (3.22).

To construct the likelihood function, let us consider:

if Y=0 then $Y_0 = 1, Y_1 = 0, Y_2 = 0, ..., Y_{K-1} = 0$

if Y=1 then $Y_0 = 0, Y_1 = 1, Y_2 = 0, ..., Y_{K-1} = 0$

if Y=2 then $Y_0 = 0, Y_1 = 0, Y_2 = 1, ..., Y_{K-1} = 0$.

if Y=K-1 then $Y_0 = 0, Y_1 = 0, Y_2 = 0, ..., Y_{K-1} = 1$.

The conditional likelihood function for a sample of n independent observations is

$$L(\beta) = \prod_{i=1}^{n} \left\{ [\pi_0(X_i)]^{Y_{0i}} [\pi_1(X_i)]^{Y_{1i}} [\pi_2(X_i)]^{Y_{2i}} ... [\pi_{K-1}(X_i)]^{Y_{K-1,i}} \right\}. \quad (3.29)$$

Taking log of (3.29) and using the fact that $\sum_{j=0}^{K-1} Y_{ji} = 1$ for each i, the log-likelihood function is

$$\ln L(\beta) = \sum_{i=1}^{n} \left\{ \left[Y_{1i} g_1(X_i) + ... + Y_{K-1,i} g_{K-1}(X_i) \right] - \left\{ \ln(1 + e^{g_1(X_i)} + ... + e^{g_K(X_i)}) \right\} \right\}. \quad (3.30)$$

Thus differentiating (3.30) and setting the derivatives equal to 0, the likelihood equations are

$$\frac{\partial \ln L(\beta)}{\partial \beta_{jk}} = \sum_{i=1}^{n} X_{ki} (Y_{ji} - \pi_{ji}) = 0 \quad (3.31)$$

where $\pi_{ji} = \pi_j(X_i)$, j=1,2,...,K-1, k=0,1,.....,p, and $X_{0i} = 1$ for each subject.

Solutions of (3.31) give the MLE's of regression parameter. The general form of the elements in the matrix of second order partial derivatives is as follows:

$$\frac{\partial^2 \ln L(\beta)}{\partial \beta_{jk} \partial \beta_{jk'}} = -\sum_{i=1}^{n} X_{k'i} X_{ki} \pi_{ji} (1 - \pi_{ji}), \quad (3.32)$$

and

$$\frac{\partial^2 \ln L(\beta)}{\partial \beta_{jk} \partial \beta_{j'k'}} = \sum_{i=1}^{n} X_{k'i} X_{ki} \pi_{ji} \pi_{j'i}, \quad (3.33)$$

where $j, j' = 1, 2, ..., K-1$ and $k, k' = 0, 1, ..., p$.

The information matrix, $I(\beta)$, is the (K-1)(p+1) by (K-1)(p+1) matrix whose elements are the negative of the expected values of the expressions given in equations (3.32) and (3.33). The asymptotic covariance matrix of the MLE is the inverse of the information matrix

$$\Sigma(\beta) = [I(\beta)]^{-1}. \quad (3.34)$$

Expression (3.34) is the same as (3.25). A more concise representation for the estimator of the information matrix, similar to (3.26), may be obtained if we express it in a form similar to the binary outcome case. Let the matrix X be the n x (p+1) matrix containing the covariates for each subject, and let the matrix V_j be the n x n diagonal matrix with general element $\hat{\pi}_{ji}(1-\hat{\pi}_{ji})$, j=1,2 ,…,K-1 and i=1,2,…,n, then the information matrix is

$$\hat{I}_j(\hat{\beta}) = X'V_jX \quad, j=1,2,...,K-1.$$

3.4.5 Measuring the Goodness of Fit

The log likelihood function for a model specified by certain link function of the generalized linear model is

$$l(y,\mu_i) = \sum_{i=1}^{n} \left[\{y_i\theta_i - b(\theta_i)\}/a(\phi) + c(y_i,\phi)\right]$$

where μ is the vector of expected value of Y and θ is the vector of canonical parameters. The log likelihood for the saturated model is

$$l(y,y) = l(y,\phi;y) = \{y\theta(y) - b(\theta(y))\}/a(\phi) + c(y,\phi)$$

$$= \sum_{i=1}^{n} \left[\{y_i\theta_i(y) - b(\theta_i(y))\}/a(\phi) + c(y_i,\phi)\right].$$

Then the deviance measure for the proposed model is

$$D(Y,\mu) = -2[l(y,\mu) - l(y,y)] = 2[l(y,y) - l(y,\mu)]. \tag{3.35}$$

The scaled deviance is defined as $D^*(y,\mu) = D(y,\mu)/\phi$.

For a binomial distribution for proportion (see Agresti, 2002), the generalized linear model parameters are:

$\theta = \ln[\pi/(1-\pi)]$, $\pi = e^{\theta}/(1+e^{\theta})$, $b(\theta) = -\ln(1-\pi)$, $a(\phi) = 1/n$

$c(y,\phi) = \ln\begin{pmatrix} n \\ y \end{pmatrix}$. The estimates of the parameters for the given model are:

$\hat{\theta} = \ln[\hat{\pi}/(1-\hat{\pi})]$, $\hat{\pi} = e^{\hat{\theta}}/(1+e^{\hat{\theta}})$, $b(\hat{\theta}) = -\ln(1-\hat{\pi})$, $a(\phi) = 1/n$,

$c(y,\hat{\phi}) = \ln\begin{pmatrix} n \\ y \end{pmatrix}$. Similarly, the estimates for the saturated model

are: $\tilde{\theta} = \ln[p/(1-p)]$, $p = e^{\tilde{\theta}}/(1+e^{\tilde{\theta}})$, $b(\tilde{\theta}) = -\ln(1-p)$, $a(\phi) = 1/n$,

$c(y,\hat{\phi}) = \ln\begin{pmatrix} n \\ y \end{pmatrix}$, where p is the sample proportion, y/n, or y=np. From (3.35) the deviance measure for grouped data is:

$$D(Y,\mu) = -2[l(y,\mu) - l(y,y)] = 2\sum_{i=1}^{k} n_i p_i \ln\frac{n_i p_i}{n_i\hat{\pi}_i} + 2\sum_{i=1}^{k}(n_i - n_i p_i)\ln\frac{n_i - n_i p_i}{n_i - n_i\hat{\pi}_i}. \tag{3.36}$$

From (3.35) the deviance measure for ungrouped data using the Bernoulli distribution can be shown as:

$$D(Y,\mu) = -2[l(y,\mu) - l(y,y)] = 2\sum_{i=1}^{n} y_i \ln\frac{y_i}{\hat{\pi}_i} + 2\sum_{i=1}^{k}(1-y_i)\ln\frac{1-y_i}{1-\hat{\pi}_i}. \quad (3.37)$$

Expression (3.37) is more useful than (3.36). It is noteworthy that usually we deal with the grouped data for a logistic regression model

$$\eta_i = g[E(y_i)] = g(\mu_i) = \ln\left[\frac{\mu_i}{1-\mu_i}\right] = \ln\left[\frac{\pi_i}{1-\pi_i}\right] = x_i'\beta = \theta_i$$

where x_i' is a row vector of (p+1) variables. If we consider that there are k distinct sets of X values, then $\sum_{i=1}^{k} n_i = n$, where n_i is the number of individuals with $X = x_i$, and y_i is the responses with Y=1 (success) out of n_i individuals in the ith group and $(n_i - y_i)$ is the number of responses with Y=0 (failures) out of n_i individuals in the same group.

Another measure of goodness of fit is based on the traditional Pearson chi-square. For the grouped data, as stated above, the estimate of y_i can be shown as follows:

$$\hat{y}_i = n_i\hat{\pi}_i = n_i e^{X'\hat{\beta}_i} / (1 + e^{X'\hat{\beta}_i}).$$

We can show the Pearson chi-square as

$$\chi^2 = \sum_{i=1}^{k} \frac{(y_i - n_i\hat{\pi}_i)^2}{n_i\hat{\pi}_i(1-\hat{\pi}_i)}. \quad (3.38)$$

The Pearson statistic given by (3.38) is approximately distributed as chi-square with n-p-1 degrees of freedom for large samples.

The scaled Pearson's chi-square is

$$\chi^2_{scaled} = \chi^2 / \phi. \quad (3.39)$$

Both (3.38) and (3.39) under certain regularity conditions, have a limiting chi-square distribution, with degrees of freedom equal to the number of observations minus the number of parameters estimated. The scaled version can be used as an approximate guide to the goodness of fit of a given model.

3.4.6 Residual Analysis in the GLM

To check the adequacy of a model or to check the model fitting, residuals can provide important information. The residual is defined as the difference between the observed and fitted values as shown below:

$$e_i = y_i - \hat{y}_i = y_i - \hat{\mu}_i .$$

The ith deviance residual in the GLM is defined as

$$r_{D_i} = \sqrt{d_i}\, sign(y_i - \hat{y}_i)$$

where d_i is the contribution of the ith observation to the deviance. For the case of logistic regression (a GLM with binomial errors and the logit link), we can show that

$$d_i = y_i \ln\left(\frac{y_i}{n_i\hat{\pi}_i}\right) + (n_i - y_i)\left[\frac{1-(y_i/n_i)}{1-\hat{\pi}_i}\right], \text{ i= 1,2,....,n}$$

where

$$\hat{\pi}_i = e^{X\beta}/(1+e^{X\beta}).$$

3.5 Logistic Regression for Dependent Binary Outcomes

We can extend the logistic regression models for serially dependent binary variables emerging from repeated measures data. The conditional probabilities for the i-th dependent variable can be expressed as a function of covariate as well as previous outcome variables. Let $Y = (Y_1, Y_2, ..., Y_n)$ be a set of n dependent binary variables, where each Y_i can take values 0 or 1, and the corresponding covariate vector is X for each Y. Then the probability of Y given X can be shown as follows:

$$P(Y|X) = P(Y_1, Y_2, ..., Y_n|X) = P(Y_1|X)P(Y_2|Y_1, X)...P(Y_n|Y_1, Y_2, ...Y_{n-1}, X). \quad (3.40)$$

We can express (3.40) in the following form for the i-th dependent variable

$$P(Y_i \mid X) = P(Y_i|Y_1, Y_2, ...Y_{i-1}, X_i) \quad (3.41)$$

and from (3.41) the probability of occurrence of the event for the i-th dependent variable is defined as:

$$P(Y_i = 1|Y_1, Y_2, ...Y_{i-1}, X) = \frac{e^{\alpha+\gamma_1 Z_{i1}+...+\gamma_{i-1}Z_{i,i-1}+\beta X_i}}{1+e^{\alpha+\gamma_1 Z_{i1}+...+\gamma_{i-1}Z_{i,i-1}+\beta X_i}},$$
$$P(Y_i = 0|Y_1, Y_2, ...Y_{i-1}, X) = \frac{1}{1+e^{\alpha+\gamma_1 Z_{i1}+...+\gamma_{i-1}Z_{i,i-1}+\beta X_i}}. \quad (3.42)$$

where $Z_{ij} = Z_{ij}(Y_{ij})$, linear functions of the Y's.

Using (3.42) the i-th logit is then

$$\theta_i = \ln\left[\frac{P(Y_i = 1|Y_1, Y_2, ...Y_{i-1}, X_i)}{P(Y_i = 0|Y_1, Y_2, ...Y_{i-1}, X_i)}\right] = \alpha + \gamma_1 Z_{i1} + ... + \gamma_{i-1}Z_{i,i-1} + \beta X_i \quad (3.43)$$

and the likelihood function for the n dependent observations is

$$L = P(Y|X) = \prod_{i=1}^{n} \frac{e^{\theta_i Y_i}}{1+e^{\theta_i}}. \quad (3.44)$$

where θ_i is given by (3.43). Then differentiating lnL, where L is given by (3.44), with respect to the parameters and equating the derivatives to zero, we can obtain the estimates.

3.6 APPLICATIONS

For applications in this chapter, we have employed the HRS data explained in chapter 1. We have considered the mobility index for the elderly population for demonstrating the logistic regression model. For no difficulty in the five tasks the index is 0 and for at least one difficulty the index is 1 or higher. To demonstrate a two-state Markov chain model, we have considered 0 for no difficulty and 1 for difficulty. Table 2.1 displayed the pooled number of transitions during two consecutive follow-ups in the period 1992-2004.

Table 3.1 below displays the fitting of a logistic regression model with single covariate age. It appears from Table 3.1 that age is positively associated with odds of difficulty in mobility ($p<0.001$). The likelihood ratio test shows that the model is significant (likelihood ratio chi square is 25.4938 and p value is less than 0.001). The Wald chi- square test shows the significance of age, where age is a continuous variable.

Table 3.1. Logistic Regression Model Fit Statistics for Mobility Index from 1992 Survey Data (Two States)

Criterion	Intercept Only	Intercept and Covariates
AIC	11693.869	11670.375
SC	11701.054	11684.745
-2 Log L	11691.869	11666.375

Test	Chi-Square	DF	p-value
Likelihood Ratio	25.4938	1	0.0001
Score	25.4983	1	0.0001
Wald	25.4459	1	0.0001

Parameter	DF	Estimate	Standard Error	Wald Chi-Square	p-value	Odds Ratio	95% Wald Confidence Limits	
Intercept	1	-2.8799	0.39170	54.0545	0.0001			
Age (years)	1	0.0354	0.00702	25.4459	0.0001	1.036	1.022	1.050

Multiple Logistic Regression

Table 3.2 which follows is an application of the logistic regression model for several selected covariates such as age, gender and race (White, Black and other races). In this example, age is a continuous variable, gender is a binary variable (male=1, and female=0), and race is a variable with three categories, hence, two design variables are defined (White, Black, other race is considered as the reference category). The results show that likelihood ratio chi-square is 248.5296 with degrees of freedom=4 (p-value<0.001). The Wald test results show positive association of age (p-value<0.001), and negative association of gender (p-value<0.001) and White race as compared to other races (p-value<0.05). However, Black race does not show any significant association with odds of difficulty in mobility among elderly as compared to that of other races (p-value>0.10).

Table 3.2. Multiple Logistic Regression Model Fit Statistics for Mobility Index (Two States) from 1992 Survey

Criterion	Intercept Only	Intercept and Covariates
AIC	11693.869	11453.339
SC	11701.054	11489.265
-2 Log L	11691.869	11443.339

Test	Chi-Square	DF	p-value
Likelihood Ratio	248.5296	4	0.0001
Score	247.4332	4	0.0001
Wald	241.7884	4	0.0001

Parameter	DF	Estimate	Standard Error	Wald Chi-Square	p-value	Odds Ratio	95% Wald Confidence Limits	
Intercept	1	-2.5512	0.4101	38.6928	0.0001			
Age (years)	1	0.0370	0.00710	27.0693	0.0001	1.038	1.023	1.052
Gender	1	-0.5797	0.0462	157.6435	0.0001	0.560	0.512	0.613
White	1	-0.2473	0.1170	4.4661	0.0346	0.781	0.621	0.982
Black	1	0.1718	0.1250	1.8872	0.1695	1.187	0.929	1.517

Table 3.3. Maximum Likelihood Ratio Test for Mobility Index from 1992 Survey (Two States)

Source	DF	Chi-Square	p-value
Intercept	2	73.350	.000
Age (years)	2	39.824	.000
Gender	2	162.487	.000
White	2	5.921	.052
Black	2	1.907	.385
Final Model	8	270.433	0.000

Parameter	Estimate	Standard Error	Chi-Square	p-value	Odds Ratio	95% Wald Confidence Limits	
For Mobility Index 1 Vs 0							
Intercept	-2.144	0.540	15.742	0.000			
Age (years)	0.015	0.009	2.671	0.102	1.015	0.997	1.034
Gender	-0.630	0.062	104.573	0.000	0.533	0.472	0.601
White	-0.120	0.158	0.574	0.449	0.887	0.650	1.210
Black	0.170	0.169	1.008	0.315	1.185	0.851	1.651
For Mobility Index 2+ Vs 0							
Intercept	-4.301	0.530	65.906	0.000			
Age (years)	0.058	0.009	39.354	0.000	1.059	1.040	1.078
Gender	-0.532	0.060	79.448	0.000	0.587	0.522	0.660
White	-0.362	0.145	6.190	0.013	0.696	0.524	0.926
Black	0.173	0.155	1.252	0.263	1.189	0.878	1.609

Multinomial Logistic Regression

The polytomous logistic regression results are summarized in Table 3.3. The outcome variable is defined as: Y=0, for no difficulty, Y=1, for little difficulty, and Y=2, for severe difficulty. We have shown here two logistic regression models, first one for mobility index, Y =1 versus Y=0, and the second one is, Y=2 versus Y=0. First part of Table 3.3 shows the test for fit of the model. The likelihood ratio chi-square is 270.433 with degrees of freedom = 8 and the p-value is less than 0.001. The model for little difficulty versus no difficulty shows that age and race variables are not significantly associated with the odds of little difficulty (p-value>0.10) but gender shows significant negative association (p-value<0.001).

Multiple Logistic Regression: Follow-up Data

The results displayed in Table 3.2 and Table 3.3 are based on cross-sectional data for a particular year. These results can be obtained from the follow-up data as well using any two consecutive follow-ups. Tables 3.4 and 3.5 summarize the results using the same outcome variables and covariates based on the data from the 1992 and 1994 follow-ups. Now, if we compare the results, it is evident that the Table 3.2 and Table 3.4 show similar significance but the estimates are quite different now. Similarly, if we compare the results presented in Table 3.3 and Table 3.5, we observe that Table 3.5 results are somewhat different in terms of the estimates. Table 3.4 and Table 3.5 provide the basis for transitional model estimation which will be further illustrated later.

Table 3.4a. Multiple Logistic Regression Model Fit Statistics for Mobility Index from 1992 & 1994 Survey (Two States)

Criterion	Intercept Only	Intercept and Covariates
AIC	23159.908	22555.806
SC	23167.731	22594.922
-2 Log L	23157.908	22545.806

Table 3.4b. Testing Global Null Hypothesis: β=0 for Mobility Index from 1992 & 1994 Survey (Two States)

Test	Chi-Square	DF	p-value
Likelihood Ratio	612.1020	4	0.0001
Score	605.6851	4	0.0001
Wald	589.2015	4	0.0001

Table 3.4c. Maximum Likelihood Estimates of Multiple Logistic Regression for Mobility Index from 1992 & 1994 Survey (Two States)

Parameter	DF	Estimate	Standard Error	Wald Chi-Square	p-value	Odds Ratio	95% Wald Confidence Limits	
Intercept	1	-3.2130	0.2841	127.9135	0.0001			
Age (years)	1	0.0518	0.0048	114.8732	0.0001	1.053	1.043	1.063
Gender	1	-0.6368	0.0326	380.8945	0.0001	0.529	0.496	0.564
White	1	-0.2644	0.0843	9.8489	0.0017	0.768	0.651	0.905
Black	1	0.1274	0.0901	1.9988	0.1574	1.136	0.952	1.355

Multinomial Logistic Regression

Table 3.5a. Maximum Likelihood Ratio Test for Mobility Index from 1992 & 1994 Survey (Three States)

Source	DF	Chi-Square	p-value
Intercept	2	200.603	.000
Age (years)	2	127.234	.000
Gender	2	392.683	.000
White	2	15.833	.000
Black	2	2.513	.285
Final Model	8	674.611	.000

Table 3.5b. Maximum Likelihood Estimates of Multinomial Logistic Regression for Mobility Index from 1992 & 1994 Survey (Three States)

Parameter	Estimate	Standard Error	Chi-Square	p-value	Odds Ratio	95% Wald Confidence Limits	
For Mobility Index 1 Vs 0							
Intercept	-3.254	0.365	79.356	0.000			
Age (years)	0.039	0.006	38.754	0.000	1.039	1.027	1.052
Gender	-0.685	0.042	260.717	0.000	0.504	0.464	0.548
White	-3.254	0.365	79.356	0.402	0.910	0.731	1.134

Table 3.5b. (Continued)

Black	0.039	0.006	38.754	0.560	1.072	0.848	1.357
For Mobility Index 2+ Vs 0							
Intercept	-0.685	0.042	260.717	-0.685	0.042	260.717	-0.685
Age (years)	-0.094	0.112	0.702	-0.094	0.112	0.702	-0.094
Gender	0.070	0.120	0.339	0.070	0.120	.339	0.070
White	-0.427	0.104	16.945	-0.427	0.104	16.945	-0.427
Black	0.171	0.110	2.423	0.171	0.110	2.423	0.171

Logistic Regression By Taking Previous Outcomes asCovariates

Table 3.6 shows an example of dependent binary outcomes. on the basis of two consecutive follow-up data from 1992, 1994 and 1996 data.. The first part of Table 3.6 (3.6a-3.6c) is based on 1992-94 data which is same as the results presented in Table 3.4a to table 3.4c. The second part of Table 3.6 is based on the follow-up data from 1992, and 1994. The covariates are same as other tables but the second part includes an additional variable Y1 which represents the transition status during the period 1992-94. The outcome variable for the second part of Table 3.6 (Table 3.6e) is defined from the transitions that occurred during 1992-94 (Y1) and in the model for first order transition, Y1 is considered as a covariate. The second part of Table 3.6f shows that age is positively (p-value<0.001) and gender is negatively associated. (P-value <0.001) for the transition from no difficulty to difficulty. In addition, we observe that the first order transition increases the risk of transition to difficulty at the second order (p-value<0.001).

Table 3.6a. Multiple Logistic Regression Model Fit Statistics for Mobility Index from 1992 & 1994 Survey (Two States)

Criterion	Intercept Only	Intercept and Covariates
AIC	23159.908	22555.806
SC	23167.731	22594.922
-2 Log L	23157.908	22545.806

Table 3.6b. Testing Global Null Hypothesis: □=0 for Mobility Index from 1992 & 1994 Survey (Three States)

Test	Chi-Square	DF	p-value
Likelihood Ratio	612.1020	4	0.0001
Score	605.6851	4	0.0001
Wald	589.2015	4	0.0001

Table 3.6c. Maximum Likelihood Estimates of Multiple Logistic Regression for Mobility Index from 1992 & 1994 Survey

Parameter	Estimate	Standard Error	Wald Chi-Square	p-value	Odds Ratio	95% Wald Confidence Limits	
Intercept	-3.2130	0.2841	127.914	0.0001			
Age (years)	0.0518	0.0048	114.873	0.0001	1.053	1.043	1.063
Gender	-0.6368	0.0326	380.895	0.0001	0.529	0.496	0.564
White	-0.2644	0.0843	9.849	0.0017	0.768	0.651	0.905
Black	0.1274	0.0901	1.999	0.1574	1.136	0.952	1.355

Table 3.6d. Multiple Logistic Regression Model Fit Statistics for Mobility Index from 1994 Survey by Taking 1992 Outcome as Covariate

Criterion	Intercept Only	Intercept and Covariates
AIC	11360.260	8927.680
SC	11367.332	8970.108
-2 Log L	11358.260	8915.680

Table 3.6e. Testing Global Null Hypothesis: □=0 for Mobility Index from 1994 Survey by Taking 1992 Outcome as Covariate

Test	Chi-Square	DF	p-value
Likelihood Ratio	2442.5805	5	0.0001
Score	2419.1958	5	0.0001
Wald	1958.6706	5	0.0001

Table 3.6f. Maximum Likelihood Estimates of Multiple Logistic Regression for Mobility Index from 1994 Survey by Taking 1992 Outcome as Covariate

Parameter	Estimate	Standard Error	Wald Chi-Square	p-value	Odds Ratio	95% Wald Confidence Limits	
Intercept	-3.2243	0.4924	42.8713	0.0001			
Age (years)	0.0396	0.00825	23.0741	0.0001	1.040	1.024	1.057
Gender	-0.5453	0.0533	104.8265	0.0001	0.580	0.522	0.643
White	-0.1935	0.1425	1.8452	0.1743	0.824	0.623	1.089
Black	0.0144	0.1527	0.0089	0.9247	1.015	0.752	1.369
Y_1	2.4280	0.0572	1804.48	0.0001	11.336	10.134	12.679

3.7. Summary

The generalized linear models are discussed in this chapter with special attention to the logistic regression model. The logit link function will be employed in subsequent chapters very extensively, so this chapter is very useful to the readers for their understanding of the remaining chapters. The logistic regression model is shown for single covariate and multiple covariates as well as for binary outcomes and polytomous outcomes. Hosmer and Lemeshow (2000) provides an excellent review of the techniques related to the logistic regression models, and McCullagh and Nelder (2000) provides the necessary details for the generalized linear models. The models are fitted in the applications section to real life data.

Chapter 4

COVARIATE DEPENDENT TWO STATE FIRST ORDER MARKOV MODEL

4.1 INTRODUCTION

This chapter introduces the covariate dependent first order Markov model with two states. The first order Markov model is represented by a 2×2 transition probability matrix. The row sums of the transition probability matrix are equal to 1. If we assume a binary outcome for each row, then the underlying distributions are assumed to be Bernoulli (for a single experiment) or binomial (for multiple experiments). In many fields of research, we observe such outcomes. Some of the experiments are: (i) occurrence of daily rainfall in consecutive days where the outcomes are denoted by yes for rainfall and no for no rainfall, (ii) incidence of disease in consecutive follow-ups in a study subject where the outcome is either occurrence of disease or no disease, (iii) change in employment status during consecutive time intervals indicating employment status (employed or unemployed), (iv) status of a machine in subsequent time points such as machine is inactive or active, (v) increase or decrease in share market index in consecutive days, (vi) increase or decline in GDP in consecutive years, etc. The common characteristic in all these examples is that the outcomes are binary and the binary outcomes are repeated over time. In other words, we obtain repeated observations over a period of time for all the individuals/subjects/items. For the sake of brevity, we will term the units as individuals.

The transition probabilities have been mainly employed to understand the underlying mechanism or pattern in the repeated observations. Recently, there have been some developments in relating transition probabilities with covariates so that the variables that influence the transitions can also be identified. In this chapter, a first order transition model is described with examples.

4.2 FIRST ORDER TRANSITION MODEL

Let us consider an experiment for a specified time period for a sample of size n. In the experiment, we have data from several follow-ups. At each follow-up, each of n units is observed. The n units in the sample produce data on the dependent variable, Y, and a covariate vector, $X' = (1, X_1, X_2, ..., X_p)$. Let us assume that the dependent variable Y can take two values, 0 and 1. If Y=1, then it indicates the occurrence of an event, otherwise Y=0. For the i-th unit at the j-th follow-up, the observations can be expressed as follows:

$Y_{ij} = 1$, if the i-th unit experiences the event of interest at the j-th follow-up,
= 0, otherwise;

$X_{ijq} = x_{ijq}$, value of the covariate X_q for the i-th unit at the j-th follow-up.

Let us consider a single stationary process $(Y_{i1}, Y_{i2}, ..., Y_{ij})$ representing the past and present responses for subject i (i= 1,.2,…,n) at follow-up j (j=1,2,…, n_i). Y_{ij} is the response at time t_{ij}. We can think of Y_{ij} as an explicit function of past history of subject i at follow-up j denoted by $H_{ij} = \{Y_{ik}, \text{k=1,2,...,j-1}\}$. The transition models for which the conditional distribution of Y_{ij} given H_{ij} depends on r prior observations, $Y_{ij-1}, ..., Y_{ij-r}$, is considered as the model of order r. Then the first order Markov model can be expressed as

$$P(Y_{ij} | H_{ij}) = P(Y_{ij} | Y_{ij-1})$$

and the corresponding transition probability matrix is given as follows:

$$\begin{array}{cc} Y_{ij-1} & Y_{ij} \\ & \begin{array}{cc} 0 & 1 \end{array} \\ \begin{array}{c} 0 \\ 1 \end{array} & \begin{bmatrix} \pi_{00} & \pi_{01} \\ \pi_{10} & \pi_{11} \end{bmatrix}. \end{array}$$

The probability of a transition from 0 at time t_{j-1} to 1 at time t_j is $\pi_{01} = P(Y_{ij} = 1 | Y_{ij-1} = 0)$ and similarly the probability of a transition from 1 at time t_{j-1} to 1 at time t_j is $\pi_{11} = P(Y_{ij} = 1 | Y_{ij-1} = 1)$. For covariate dependence, let us define the following notations: $X'_{ij-1} = [1, X_{ij-1,1}, ..., X_{ij-1,p}]$ = vector of covariates for the i-th person at the (j-1)th follow-up;

$\beta'_0 = [\beta_{00}, \beta_{01}, ..., \beta_{0p}]$ = vector of parameters for the transition from 0,

$\beta'_1 = [\beta_{10}, \beta_{11}, ..., \beta_{1p}]$ = vector of parameters for the transition from 1.

Then the transition probabilities can be defined in terms of functions of the covariates as

$$\pi_{s1} = P(Y_{ij} = 1 | Y_{ij-1} = s, X_{ij-1}) = e^{\beta'_s X_{ij-1}} / (1 + e^{\beta'_s X_{ij-1}}), \text{ (s=0,1).} \quad (4.1)$$

4.3 Likelihood Function

Then using (4.1) the likelihood function can be defined as

$$L = \prod_{s=0}^{1}\prod_{m=0}^{1}\prod_{i=1}^{n}\prod_{j=1}^{n_i}\left[\{\pi_{sm}\}^{\delta_{smij}}\right] \qquad (4.2)$$

where n_i = total number of follow-up observations since the entry into the study for the i-th individual; δ_{smij}=1 if a transition type s-m is observed during j-th follow-up for the i-th individual. Then taking log of (4.2) we can express the log likelihood function as:

$$\ln L = \ln L_0 + \ln L_1$$

where L_0 and L_1 correspond to s = 0 and s = 1 respectively.

Hence,

$$\ln L_0 = \sum_{i=1}^{n}\sum_{j=1}^{n_i}\left[\delta_{01ij}\{\beta_{01}'X_{ij-1}\} - (\delta_{00ij}+\delta_{01ij})\ln\{1+e^{\beta_{01}'X_{ij-1}}\}\right]$$

and

$$\ln L_1 = \sum_{i=1}^{n}\sum_{j=1}^{n_i}\left[\delta_{11ij}\{\beta_{11}'X_{ij-1}\} - (\delta_{10ij}+\delta_{11ij})\ln\{1+e^{\beta_{11}'X_{ij-1}}\}\right]. \qquad (4.3)$$

Differentiating (4.3) with respect to the parameters and solving the following equations we obtain the likelihood estimates for 2(p+1) parameters:

$$\frac{\partial \ln L_0}{\partial \beta_{01q}} = 0,\ q=1,2,\ldots,p;$$

and

$$\frac{\partial \ln L_1}{\partial \beta_{11q}} = 0,\ q=1,2,\ldots,p. \qquad (4.4)$$

The first derivatives with respect to the first set of parameters in (4.4) are

$$\frac{\partial \ln L_0}{\partial \beta_{01q}} = \sum_{i=1}^{n}\sum_{j=1}^{n_i}\left[X_{ij-1,q}\left\{\delta_{00ij} - (\delta_{00ij}+\delta_{01ij})\frac{e^{\beta_{01}'X_{ij-1}}}{1+e^{\beta_{01}'X_{ij-1}}}\right\}\right] \qquad (4.5)$$

q=0,1,2,……,p.

Similarly first derivatives with respect to the second set are

$$\frac{\partial \ln L_1}{\partial \beta_{11q}} = \sum_{i=1}^{n}\sum_{j=1}^{n_i}\left[X_{ij-1,q}\left\{\delta_{10ij} - (\delta_{10ij}+\delta_{11ij})\frac{e^{\beta_{011}'X_{ij-1}}}{1+e^{\beta_{011}'X_{ij-1}}}\right\}\right], \qquad (4.6)$$

q=0,1,2,……,p.

We can solve for the sets of parameters by equating (4.5) and (4.6) to zero.

The second derivatives are:

$$\frac{\partial^2 \ln L_0}{\partial \beta_{01q}\partial \beta_{01l}} = -\sum_{i=1}^{n}\sum_{j=1}^{n_i}\left[X_{ij-1,q}X_{ij-1,l}\left\{(\delta_{00ij}+\delta_{01ij})\pi_{00}(X_{ij-1})\pi_{01}(X_{ij-1})\right\}\right] \qquad (4.7)$$

$$\frac{\partial^2 \ln L_1}{\partial \beta_{11q} \partial \beta_{11l}} = -\sum_{i=1}^{n} \sum_{j=1}^{n_i} \left[X_{ij-1,q} X_{ij-1,l} \left\{ (\delta_{10ij} + \delta_{11ij}) \pi_{10}(X_{ij-1}) \pi_{11}(X_{ij-1}) \right\} \right]. \quad (4.8)$$

Here (4.7) and (4.8) are obtained from (4.5) and (4.6), respectively. For brevity, let us consider two types of transitions 0-1 and 1-1 and let us denote $Y_{ij} = \delta_{ij}$ separately for 0-1 and 1-1. Then, in summarized form the score functions are:

$$\sum_{i=1}^{n} \sum_{j=1}^{n_i} X_{ij-1,q}[Y_{ij} - \pi(X_{ij-1})] = 0, \text{ q=0,1,2,.....,p.}$$

These equations are applied separately for transitions 0-1 and 1-1.

The variances and covariances of the estimated coefficients can be obtained from:

$$I_{ql}^{*} = -\sum_{i=1}^{n} X_{ij-1,q} X_{ij-1,l} \pi_{ij-1}(1 - \pi_{ij-1})$$

for q,l=0,1,…,p where $\pi_{ij-1} = \pi(X_{ij-1})$. Let $I_{ql} = (-1) I_{ql}^{*}$. Then I is the information matrix.

4.4 Testing for the Significance of Parameters

The vectors of 2 sets of parameters for the first order Markov model can be represented by the following vector:

$$\beta = [\beta_1', \beta_2']', \quad \beta_0 = [\beta_{10}, \beta_{20}]'$$

where $\beta_w' = \left[\beta_{w0}, \beta_{w1}, \ldots\ldots, \beta_{wp} \right]$, w=1,2.

To test the null hypothesis $H_0 : \beta_{11} = \ldots = \beta_{1p} = \beta_{21} = \ldots = \beta_{2p} = 0$, we can employ the usual likelihood ratio test

$$-2[\ln L(\beta_0) - \ln L(\beta)] \approx \chi^2_{2p}.$$

To test the significance of the q-th parameter (q=0,1,2,…,p) of the w-th (w=1,2) set of parameters, the null hypothesis is $H_0 : \beta_{wq} = 0$ and the corresponding Wald test is

$$W = \hat{\beta}_{wq} / se(\hat{\beta}_{wq}).$$

4.5 Application

In this chapter also we have used the HRS data (see Section 1.2.1.1). For a two state Markov chain model, we have considered 0 for no difficulty and 1 for difficulty. Table 4.1 displays the pooled number of transitions during two consecutive follow-ups in the period 1992-2004. It is evident from Table 4.1 that the probabilities of staying in the no difficulty and difficulty states during consecutive follow-ups are 0.800 and 0.772, respectively. The transition probability from no difficulty to difficulty is 0.200 and from difficulty to no difficulty is 0.228 respectively.

Table 4.1. Transition Counts and Transition Probabilities for First Order Two State Mobility Index

States	Transition Count		Transition Probability		Total Count
	0	**1**	**0**	**1**	
0	22461	5621	0.800	0.200	28082
1	3733	12636	0.228	0.772	16369

To test for specified values of the transition probabilities as shown in Table 4.2, we can use the chi square test as described below.

The null hypothesis is $H_0 : \pi_{sm} = \pi_{sm}^0$ and the test statistic is

$$\chi^2 = \sum_{s=0}^{1} \sum_{m=0}^{1} \frac{n_{s.}(\hat{\pi}_{sm} - \pi_{sm}^0)^2}{\pi_{sm}^0}$$

which is χ^2 with 2(2-1)-d degrees of freedom under H_0 where d is the number of zeros in π_{sm}^0.

Similarly, the likelihood ratio test (LRT) is

$$-2\ln\Lambda = 2\sum_{s=0}^{1} \sum_{m=0}^{1} n_{sm} \ln\frac{n_{sm}}{n_{s.}\pi_{sm}^0}.$$

This is χ^2 with 2(2-1) degrees of freedom under H_0.

Table 4.2. Test for Inference about First Order Two State Mobility Index

Test-statistics	Value	df	p-value
Chi-square	14940.771	2	0.000
LRT	15927.360	2	0.000

We can use the two tests for stationarity that are described in Chapter 2. The first test we employ here is based on pooled data on transitions which is displayed in Table 4.3. The second one is based on the consecutive follow-ups.

(i) Test for Stationarity Based on Pooled Transition Data

The null hypothesis is $H_0 : \pi_{sm}^t = \pi_{sm}$ and the likelihood ratio chi square is

$$-2\ln\Lambda = 2\sum_{t=1}^{T} \sum_{s=0}^{1} \sum_{m=0}^{1} n_{sm}^t \ln\frac{n_{sm}^t}{n_{s.}^{t-1}\pi_{sm}}.$$

where π_{sm} represents the transition probabilities for the pooled time.

This is χ^2 with (T-1)2(2-1) degrees of freedom under H_0. Table 4.3 shows that the value of chi square for the data under study is 80.558 with degree of freedom 12 and the corresponding p value is less than 0.001. Hence, the null hypothesis of stationarity can be rejected for the given data.

(ii) Test for Stationarity Based on Consecutive Follow-up Data

The null hypothesis is $H_0 : \pi_{sm}^{t} = \pi_{sm}^{t+1}$ where

π_{sm}^{t} = transition probability from s to m for order 1 at time t and

π_{sm}^{t+1} = transition probability from s to m for for order 1 at time t+1

then the likelihood ratio test statistic is

$$-2\ln\Lambda = 2\sum_{t=1}^{T-1}\ \sum_{s=0}^{1}\sum_{m=0}^{1} n_{sm}^{t} \ln\frac{n_{sm}^{t}}{n_{s.}^{t-1}\pi_{sm}^{t+1}}.$$

This is χ^2 with (T-2)2(2-1) degrees of freedom under H_0. The chi square based on the consecutive follow-up data is 67.789. The corresponding p value for chi square with 10 degrees of freedom is less than 0.001. The null hypothesis of stationarity may be rejected under this test procedure as well.

Table 4.3. Stationary Test for First Order Two State Mobility Index

Time	D.F.	Comparison between Consecutive Time Points		Comparison with Pooled Transition Matrix	
		Chi-square	p-value	Chi-square	p-value
2	2	25.314	0.000	9.851	0.007
3	2	4.167	0.384	10.744	0.030
4	2	11.152	0.084	19.119	0.004
5	2	25.092	0.002	1.077	0.998
6	2	2.065	0.996	14.613	0.147
7	2			25.154	0.014
Overall χ^2	10,12	67.789	0.000	80.558	0.000

The fitting of a first order Markov model with covariate dependence is displayed in Table 4.4. In a 2x2 transition probability matrix, we have two independent models, one for each row. Here, we fit models for the transitions from 0 to 1 and from 1 to 0. In both the cases, we have employed three variables, age, gender and race. Age is a continuous variable here, gender is a binary variable with gender=1 for male and 0 for female. Race is a categorical variable with more than two categories and we redefined race by three categories, White, Black and other races. White and Black are two design variables and the category of other races is considered as a reference category.

Table 4.4 shows that age is positively associated with transition from no difficulty to difficulty (p-value<0.001). Males have lower risk of making a transition from no difficulty to difficulty as compared to that of females. It is also observed that the whites are less likely to experience 0-1 type of transition as compared to the other races. As expected, the transition from difficulty to no difficulty is negatively associated with age and positively associated with gender. The White or Black races do not show any significant difference from that of the other races.

Table 4.4. Estimates of Parameters of Logistic Regression Model for First Order Two State Mobility Index

Variables	Estimate	SE	t-value	p-value	95% C.I. LL	95% C.I. UL
			FIRST ORDER			
			TRANSITION TYPE: $0 \to 1$			
Constant	-2.139	0.210	-10.185	0.000	-2.551	-1.728
Age	0.019	0.003	5.687	0.000	0.012	0.025
Gender	-0.511	0.030	-16.890	0.000	-0.570	-0.451
White	-0.162	0.082	-1.984	0.047	-0.322	-0.002
Black	0.150	0.088	1.700	0.089	-0.023	0.323
			TRANSITION TYPE: $1 \to 0$			
Constant	-0.186	0.268	-0.694	0.488	-0.712	0.340
Age	-0.019	0.004	-4.527	0.000	-0.027	-0.011
Gender	0.265	0.038	6.904	0.000	0.190	0.340
White	0.033	0.099	0.331	0.741	-0.161	0.227
Black	-0.107	0.106	-1.010	0.312	-0.315	0.101
Model χ^2 (df=10)			15242.43 (0.000)			
LR χ^2 (df=10)			16388.83 (0.000)			

4.6 SUMMARY

This chapter introduces the covariate dependent Markov model for first order. Two models are fitted for two state first order Markov models, one for transition type 0-1 and other one is for 1-0. It is noteworthy that instead of 1-0, we may also consider 1-1 without loss of generality. The readers can review the paper by Muenz and Rubinstein (1985). This chapter shows both the estimation and test procedures. The test procedures include: (i) test for specified transition probabilities using goodness of fit type chi square and likelihood ratio test procedures, and (ii) test for stationarity using pooled as well as consecutive follow-up counts.

Chapter 5

COVARIATE DEPENDENT TWO STATE SECOND ORDER MARKOV MODEL

5.1 INTRODUCTION

In Chapter 4, we introduced the two state covariate dependent first order model. In many occasions, first order Markov models may fail to characterize the repeated observations over time. Instead of just prior observation, prior two observations may be of importance in analyzing the output. In that case, we need to consider the second order Markov model. In a rainfall analysis, if we consider Day 1, Day 2 and Day 3 consecutively over time, then Day 1 and Day 2 outcomes can explain Day 3 outcome in a second order model. For a second order model with binary outcomes, we have to consider two 2×2 tables of transition probabilities, one each for given Day 1 values. This chapter shows the procedure to extend the first order covariate dependent model to a second order model.

5.2 SECOND ORDER MODEL

As in Chapter 4, let us consider a single stationary process $(Y_{i1}, Y_{i2}, ..., Y_{ij})$ for subject i (i= 1,.2,…,n) at follow-up j (j=1,2,…, n_i), where Y_{ij} is the response at time t_{ij}. The past history of subject i at follow-up j is denoted by $H_{ij} = \{Y_{ik},\ \text{k=1,2,...,j-1}\}$. If we assume that the transition model for conditional distribution of Y_{ij} given H_{ij} depends on 2 prior observations, Y_{ij-1}, Y_{ij-2}, then it is considered as the model of order 2. The binary outcome at time t_{ij} is defined as Y_{ij}=1, if an event occurs for the i-th subject at the j-th follow-up, Y_{ij}=0, otherwise.

Then the second order Markov model can be expressed as

$$P(Y_{ij} | H_{ij}) = P(Y_{ij} | Y_{ij-2}, Y_{ij-1}). \tag{5.1}$$

In other words, the transition model (5.1) for order 2 presents the conditional distribution of Y_{ij} given H_{ij} depending on two most recent prior observations Y_{ij-1}, Y_{ij-2} where Y_{ij-1}, Y_{ij-2} =0, 1.

The second order transition probabilities for time points t_{ij-2}, t_{ij-1} and t_{ij} at follow-up j with corresponding outcomes Y_{ij-2}, Y_{ij-1} and Y_{ij}, respectively, can be shown as follows:

Y_{ij-2}	Y_{ij-1}	Y_{ij} = 0	Y_{ij} = 1
0	0	π_{000}	π_{001}
0	1	π_{010}	π_{011}
1	0	π_{100}	π_{101}
1	1	π_{110}	π_{111}

Let $X'_{ij-2} = \left[1,\ X_{ij-2,1},\ \ldots\ldots,\ X_{ij-2,p}\right]$ = vector of covariates for the i-th person at the (j-2)-th follow-up; and let us define the parameter vectors for the second order Markov models as follows:

$\beta'_{001} = \left[\beta_{0010},\ \beta_{0011},\ \ldots,\ \beta_{001p}\right]$ = vector of parameters for the transition type 001,

$\beta'_{011} = \left[\beta_{0110},\ \beta_{0111},\ \ldots,\ \beta_{011p}\right]$ = vector of parameters for the transition type 011,

$\beta'_{101} = \left[\beta_{1010},\ \beta_{1011},\ \ldots,\ \beta_{101p}\right]$ = vector of parameters for the transition type 101,

$\beta'_{111} = \left[\beta_{1110},\ \beta_{1111},\ \ldots,\ \beta_{111p}\right]$ = vector of parameters for the transition type 111.

Using the definitions of vectors of covariates and parameters mentioned in the previous section, we can define the transition probabilities as

$$\pi_{001}(Y_{ij} = 1 | Y_{ij-2} = 0, Y_{ij-1} = 0, X_{ij-2}) = e^{\beta'_{001} X_{ij-2}} / (1 + e^{\beta'_{001} X_{ij-2}}),$$

$$\pi_{011}(Y_{ij} = 1 | Y_{ij-2} = 0, Y_{ij-1} = 1, X_{ij-2}) = e^{\beta'_{011} X_{ij-2}} / (1 + e^{\beta'_{011} X_{ij-2}}),$$

$$\pi_{101}(Y_{ij} = 1 | Y_{ij-2} = 1, Y_{ij-1} = 0, X_{ij-2}) = e^{\beta'_{101} X_{ij-2}} / (1 + e^{\beta'_{101} X_{ij-2}}),$$

$$\pi_{111}(Y_{ij} = 1 | Y_{ij-2} = 1, Y_{ij-1} = 1, X_{ij-2}) = e^{\beta'_{111} X_{ij-2}} / (1 + e^{\beta'_{111} X_{ij-2}}).$$

It may be noted here that $\pi_{000} + \pi_{001} = 1$, $\pi_{010} + \pi_{011} = 1$, $\pi_{100} + \pi_{101} = 1$, and $\pi_{110} + \pi_{111} = 1$.

5.3 Likelihood Function

The likelihood function, as a generalization of (4.2), can be expressed as
From the log-likelihood function can be shown as (5.2)

$$L=\prod_{i=1}^{n}\prod_{j=1}^{n_i}\left[\{\pi_{000}\}^{\delta_{000ij}}\{\pi_{001}\}^{\delta_{001ij}}\right]\left[\{\pi_{010}\}^{\delta_{010ij}}\{\pi_{011}\}^{\delta_{011ij}}\right]\left[\{\pi_{100}\}^{\delta_{100ij}}\{\pi_{101}\}^{\delta_{101ij}}\right]\left[\{\pi_{110}\}^{\delta_{110ij}}\{\pi_{111}\}^{\delta_{111ij}}\right]. \quad (5.2)$$

$$\ln L=\ln L_1+\ln L_2+\ln L_3+\ln L_4 . \quad (5.3)$$

In (5.3), L_1, L_2, L_3 and L_4 correspond to the terms involving transitions of types 001, 011, 101, and 111, respectively. Thus,

$$L_1=\prod_{i=1}^{n}\prod_{j=1}^{n_i}\left[\{\pi_{000}\}^{\delta_{000ij}}\{\pi_{001}\}^{\delta_{001ij}}\right]$$

which can be expressed as

$$L_1=\prod_{i=1}^{n}\prod_{j=1}^{n_i}\left[\left\{\frac{1}{1+e^{\beta'_{001}X_{ij-2}}}\right\}^{\delta_{000ij}}\left\{\frac{e^{\beta'_{001}X_{ij-2}}}{1+e^{\beta'_{001}X_{ij-2}}}\right\}^{\delta_{001ij}}\right].$$

Taking log we obtain

$$\ln L_1=\sum_{i=1}^{n}\sum_{j=1}^{n_i}\left[\delta_{000ij}\beta'_{001}X_{ij-2}-\left(\delta_{000ij}+\delta_{001ij}\right)\left(1+e^{\beta'_{001}X_{ij-2}}\right)\right]. \quad (5.4)$$

The first derivatives of (5.4) with respect to parameters are

$$\frac{\partial \ln L_1}{\partial \beta_{001q}}=\sum_{i=1}^{n}\sum_{j=1}^{n_i}\left[X_{ij-2,q}\left\{\delta_{000ij}-(\delta_{000ij}+\delta_{001ij})\frac{e^{\beta'_{001}X_{ij-2}}}{1+e^{\beta'_{001}X_{ij-2}}}\right\}\right] \quad q=0,1,2,...,p. \quad (5.5)$$

Similarly from $\ln L_2$, $\ln L_3$ and $\ln L_4$, we can show the following :

$$\frac{\partial \ln L_2}{\partial \beta_{011q}}=\sum_{i=1}^{n}\sum_{j=1}^{n_i}\left[X_{ij-2,q}\left\{\delta_{010ij}-(\delta_{010ij}+\delta_{011ij})\frac{e^{\beta'_{011}X_{ij-2}}}{1+e^{\beta'_{011}X_{ij-2}}}\right\}\right], \quad (5.6)$$

$$\frac{\partial \ln L_3}{\partial \beta_{101q}}=\sum_{i=1}^{n}\sum_{j=1}^{n_i}\left[X_{ij-2,q}\left\{\delta_{100ij}-(\delta_{100ij}+\delta_{101ij})\frac{e^{\beta'_{101}X_{ij-2}}}{1+e^{\beta'_{101}X_{ij-2}}}\right\}\right], \quad (5.7)$$

$$\frac{\partial \ln L_4}{\partial \beta_{111q}}=\sum_{i=1}^{n}\sum_{j=1}^{n_i}\left[X_{ij-2,q}\left\{\delta_{110ij}-(\delta_{110ij}+\delta_{111ij})\frac{e^{\beta'_{111}X_{ij-2}}}{1+e^{\beta'_{111}X_{ij-2}}}\right\}\right]. \quad (5.8)$$

We can solve for the sets of parameters equating first derivatives (5.5), (5.6), (5.7) and (5.8) to zero. Each set consists of (p+1) parameters and hence the total number of parameters to be estimated here is 4(p+1).

The second derivatives are:

$$\frac{\partial^2 \ln L_1}{\partial \beta_{001q} \partial \beta_{001l}} = -\sum_{i=1}^{n}\sum_{j=1}^{n_i}\left[X_{ij-2,q}X_{ij-2,l}\left\{(\delta_{000ij}+\delta_{001ij})\pi_{000}(X_{ij-2})\pi_{001}(X_{ij-2})\right\}\right], \quad (5.9)$$

$$\frac{\partial^2 \ln L_2}{\partial \beta_{011q} \partial \beta_{011l}} = -\sum_{i=1}^{n}\sum_{j=1}^{n_i}\left[X_{ij-2,q}X_{ij-2,l}\left\{(\delta_{010ij}+\delta_{011ij})\pi_{010}(X_{ij-2})\pi_{011}(X_{ij-2})\right\}\right], \quad (5.10)$$

$$\frac{\partial^2 \ln L_3}{\partial \beta_{101q} \partial \beta_{101l}} = -\sum_{i=1}^{n}\sum_{j=1}^{n_i}\left[X_{ij-2,q}X_{ij-2,l}\left\{(\delta_{100ij}+\delta_{101ij})\pi_{100}(X_{ij-2})\pi_{101}(X_{ij-2})\right\}\right], \quad (5.11)$$

$$\frac{\partial^2 \ln L_4}{\partial \beta_{111q} \partial \beta_{111l}} = -\sum_{i=1}^{n}\sum_{j=1}^{n_i}\left[X_{ij-2,q}X_{ij-2,l}\left\{(\delta_{110ij}+\delta_{111ij})\pi_{110}(X_{ij-2})\pi_{111}(X_{ij-2})\right\}\right]. \quad (5.12)$$

Inverse of the (-1)(second derivative) provide estimates of the variance-covariance for the respective estimates of the parameters. The second derivatives (5.9), (5.10), (5.11) and (5.12) are obtained from derivatives (5.5), (5.6), (5.7) and (5.8), respectively.

5.4 Testing for the Significance of Parameters

The vectors of 2^2 (=4) sets of parameters for the second order Markov model can be represented by the following vector:

$$\beta = [\beta'_{001}, \beta'_{011}, \beta'_{101}, \beta'_{111}]' \text{ and } \beta_0 = [\beta_{0010}, \beta_{0110}, \beta_{1010}, \beta_{1110}]'$$

To test the null hypothesis H_0 of joint vanishing of all the regression parameters (excluding the intercepts), we can employ the usual likelihood ratio test

$$-2[\ln L(\beta_0) - \ln L(\beta)] \approx \chi^2_{2^2 p}.$$

To test the significance of the qth (q=0,1,…,p) parameter of the w-th (w=1,2,…, 2^2) set of parameters, the null hypothesis is $H_0 : \beta_{wq} = 0$ and the corresponding Wald test is

$$W = \hat{\beta}_{wq} / se(\hat{\beta}_{wq}).$$

5.5 Applications

For applications in this chapter, we have employed the HRS mobility index data (section 1.2.1 D.1). We have considered data for the period 1992-2004. As mentioned in previous chapters, in all the waves, mobility index for the elderly population was constructed on the basis of five tasks including walking several blocks, walking one block, walking across the room, climbing several flights of stairs, climbing one flight of stairs. For no difficulty in the five tasks the index is 0 and for at least one difficulty the index is 1 or higher. Let us consider two states, 0 for no difficulty and 1 for difficulty. Table 4.1 displays the pooled number of transitions during two consecutive follow-ups in the period 1992-2004.

Table 5.1 reveals that the probability of remaining in no difficulty mobility index (0) is 0.847 and probability of making transition from previous two consecutive follow-ups in the no difficulty (0) and currently in difficulty in mobility index (1) is 0.153. Similarly, probability of transition types 0-1-0 and 0-1-1 are 0.413 and 0.587, and probability of transition types 1-0-0 and 1-0-1 are 0.523 and 0.477 respectively. All these estimates of transition probabilities are obtained from the pooled estimates.

Table 5.1. Transition Counts and Transition Probabilities for Second Order Two State Mobility Index

States		Transition Count		Transition Probability		Total Count
		0	1	0	1	
0	0	15687	2833	0.847	0.153	18520
0	1	1825	2589	0.413	0.587	4414
1	0	1535	1399	0.523	0.477	2934
1	1	1263	8098	0.135	0.865	9361

Now let us consider a test procedure for testing the null hypothesis $H_0 : \pi_{ijk} = \pi_{ijk}^0$ and the test statistic is

$$\chi^2 = \sum_{i=0}^{1} \sum_{j=0}^{1} \sum_{k=0}^{1} \frac{n_{ij.}(\hat{\pi}_{ijk} - \pi_{ijk}^0)^2}{\pi_{ijk}^0}$$

which is χ^2 with 2.2(2-1)-d degrees of freedom under H_0 where d is the number of zero π_{ijk}^0 s. This is a simplified version of the test procedure for higher order illustrated in Chapter 2. Similarly, the likelihood ratio test shows

$$-2\ln\Lambda = 2\sum_{i=0}^{1} \sum_{j=0}^{1} \sum_{k=0}^{1} n_{ijk} \ln \frac{n_{ijk}}{n_{ij.}\pi_{ijk}^0}.$$

This is χ^2 with 2.2(2-1) degrees of freedom under H_0. This test is simplified for order 2 based on the tests for higher order presented in Chapter 2.

Table 5.2 summarizes the results based on the chi square and likelihood ratio tests for $H_0 : \pi_{ijk} = 0.5$. The results indicate that the second order transition probabilities are not 0.5.

Table 5.2. Test for Inference about Second Order Two State Mobility Index

Test-statistics	Value	DF	p-value
Chi-square	14050.618	4	0.000
LRT	15536.706	4	0.000

Now let us test for the stationarity of transition probabilities for the second order. We have used two tests here: (i) test for stationarity on the basis of pooled transition data, and (ii) test for stationarity based on consecutive follow-up data.

(iii) Test for Stationarity Based on Pooled Transition Data

The null hypothesis is $H_0 : \pi_{ijk}^{t} = \pi_{ijk}$ and the likelihood ratio chi square is

$$-2\ln\Lambda = 2\sum_{t=1}^{7-2}\ \sum_{i=0}^{1}\sum_{j=0}^{1}\sum_{k=0}^{1} n_{ijk}^{t} \ln \frac{n_{ijk}^{t}}{n_{ij.}^{t-1}\,\pi_{ijk}} .$$

where π_{ijk} represents the transition probabilities for the pooled time. The general procedure for higher order Markov models is illustrated in Chapter 2. This is χ^2 with (7-2)2.2.(2-1) degrees of freedom under H_0. Using the pooled estimate of transition probabilities as the null hypothesis values, we observe from Table 5.3 that the chi square is 61.608. The degrees of freedom are 20 and the corresponding p-value is less than 0.001. Hence, the null hypothesis of stationarity for the second order transition probability can be rejected for the given data.

(iv) Test for Stationarity Based on Consecutive Follow-up Data

This test procedure allows us to understand the process of stationarity better. Under stationarity, the consecutive transition probabilities with respect to time should remain constant. Any deviation from the constant transition probabilities at consecutive follow-up times would shed doubt regarding the stationarity of the underlying stationary process. In this case, the null hypothesis is $H_0 : \pi_{ijk}^{t} = \pi_{ijk}^{t+1}$

where

π_{ijk}^{t} = transition probability from i to j to k for order 2 at time t and

π_{ijk}^{t+1} = transition probability from i to j to k for for order 2 at time t+1.

The likelihood ratio test statistic is (see Chapter 2 for the general procedure for order r):

$$-2\ln\Lambda = 2\sum_{t=1}^{7-3}\sum_{i=0}^{1}\sum_{j=0}^{1}\sum_{k=0}^{1} n_{ijk}^{t} \ln\frac{n_{ijk}^{t}}{n_{ij.}^{t-1}\pi_{ijk}^{t+1}}.$$

The degrees of freedom for the χ^2 based on the likelihood ratio test is (7-3)2.2.(2-1) under H_0. It is evident from Table 5.3 that the value of chi square is 87.003 mostly attributable to the third and the fifth follow-up times. The degrees of freedom is 16 and the p-value is less than 0.001. In other words, we may reject the null hypothesis of stationarity for the given data on mobility index.

Table 5.3. Stationary Test for Second Order Two State Mobility Index

Time	D.F.	Comparison between Consecutive time points		Comparison with pooled transition matrix	
		Chi-square	p-value	Chi-square	p-value
3	4	30.886	0.000	4.766	0.312
4	4	12.051	0.149	20.148	0.010
5	4	40.244	0.000	8.434	0.750
6	4	3.822	0.999	15.876	0.462
7	4			12.385	0.902
Overall χ^2	16,20	87.003	0.000	61.608	0.000

Table 5.4. Estimates of Parameters of Logistic Regression Model for Second Order Two State Mobility Index

Variables	Estimate	SE	t-value	p-value	95% C.I. LL	95% C.I. UL
FIRST ORDER						
TRANSITION TYPE: 0→0→1						
Constant	-2.974	0.309	-9.612	0.000	-3.581	-2.368
Age	0.025	0.005	5.155	0.000	0.016	0.035
Gender	-0.370	0.041	-9.018	0.000	-0.450	-0.290
White	-0.091	0.116	-0.789	0.430	-0.318	0.135
Black	0.259	0.124	2.082	0.037	0.015	0.502
TRANSITION TYPE: 0→1→0						
Constant	-1.145	0.450	-2.546	0.011	-2.026	-0.264
Age	0.015	0.007	2.059	0.040	0.001	0.029
Gender	0.146	0.062	2.344	0.019	0.024	0.268
White	-0.124	0.167	-0.745	0.456	-0.451	0.202
Black	-0.212	0.179	-1.188	0.235	-0.563	0.138
TRANSITION TYPE: 1 →0→1						
Constant	-0.402	0.564	-0.713	0.476	-1.508	0.703
Age	0.008	0.009	0.925	0.355	-0.009	0.026
Gender	-0.232	0.076	-3.048	0.002	-0.381	-0.083
White	-0.092	0.196	-0.468	0.640	-0.477	0.293
Black	-0.109	0.212	-0.515	0.607	-0.523	0.306

Table 5.4. (Continued)

TRANSITION TYPE: 1→1→0						
Constant	-1.058	0.472	-2.242	0.025	-1.984	-0.133
Age	-0.016	0.007	-2.084	0.037	-0.030	-0.001
Gender	0.217	0.063	3.436	0.001	0.093	0.340
White	0.079	0.163	0.480	0.631	-0.242	0.399
Black	-0.041	0.174	-0.234	0.815	-0.382	0.301
Model χ^2 (df=10)			14157.88 (0.000)			
LRT (df=10)			15723.02 (0.000)			

The fitting of logistic regression models for the second order are presented in Table 5.4. Now we have 4 independent models as shown in the previous sections. We have used the same variables as before and the transition probabilities for which the models are fitted are: 0-0-1, 0-1-0, 1-0-1 and 1-1-0. It may be noted here that the models are based on the change in status in the last two consecutive follow-ups. In the 0-0-1 type of transition, we observe that age, gender and black race compared to other races show significant association. Both age and black race show positive associations with move to no difficulty to difficulty but gender shows negative association. On the other hand, for transition type 0-1-0, we observe that age and gender are significantly associated, both having positive associations. Only gender shows any significant association for 1-0-1 type of transition with negative association, similar to that of the 0-0-1. Finally, the transition type 1-1-0 indicates negative age and positive gender associations.

5.6 SUMMARY

This chapter extends the first order Markov model for two states introduced in Chapter 4 to second order model for two states. There are four different models for four rows of the second order transition probability matrix for two states. The estimates and tests are displayed. The tests for specified transition probabilities and stationarity are also shown for transition count matrix. The models are fitted to the HRS data on mobility index for elderly population. For higher order models, Islam and Chowdhury (2006, 2008) provide the details.

Chapter 6

Covariate Dependent Two State Higher Order Markov Model

6.1 Introduction

We have discussed the first and second order covariate dependent Markov models in Chapter 4 and Chapter 5, respectively. This chapter highlights the generalized form for higher order models. It is noteworthy that in a higher order model, the sets of parameters to be estimated will increase substantially. However, the advantage of this model lies in the fact that we can obtain a very extensive analysis on the relationships between the selected covariates and the transition probabilities that can provide useful information.

6.2 Higher Order Model

In this section, we extend the model to order r for a single stationary process $(Y_{i1}, Y_{i2}, ..., Y_{ij})$ which represents the past and present responses for subject i (i= 1,.2,…,n) at follow-up j (j=1,2,…, n_i), Y_{ij} being the response at time t_{ij}. In this case, Y_{ij} is an explicit function of past history of subject i at follow-up j denoted by $H_{ij} = \{Y_{ik}, \text{k=1,2,...,j-1}\}$. The transition models for which the conditional distribution of Y_{ij} given H_{ij} depends on r prior observations, $Y_{ij-1}, ..., Y_{ij-r}$, is considered as the model of order r. As before, we consider only binary response here. The outcomes are Y_{ij} =1, if an event occurs for the i-th subject at the j-th follow-up, Y_{ij} =0, otherwise.

Then the Markov chain of order r can be expressed as

$$P(Y_{ij} | H_{ij}) = P(Y_{ij} | Y_{ij-r}, ..., Y_{ij-1}). \qquad (6.1)$$

As (6.1) shows, all the r most recent prior observations are considered here and the conditional distribution of Y_{ij} given H_{ij} depends on r prior observations $Y_{ij-1}, Y_{ij-2}, ..., Y_{ij-r}$ where $Y_{ij-1}, Y_{ij-2}, ..., Y_{ij-r}$ =0, 1.

For the Makov model of order r, we can define $X'_{ij-r} = \left[1, \text{X}_{\text{ij-r,1}}, \text{ ...., } \text{X}_{\text{ij-r,p}}\right]$ = vector of covariates for the i-th person at the (j-r)-th follow-up; and define the parameter vectors for the r-th order Markov models as follows:

$$\beta'_m = \left[\beta_{m0}, \beta_{\text{m1}}, \text{ ...., } \beta_{\text{mp}}\right]$$ = vector of parameters for the transition type m,

where m denotes one of 2^r transition types generated from a Markovian transition probability matrix of order r.

For the r-th order model, with outcomes $y_{ij} = 0$ or 1, at time points $t_{i,j-r}, ..., t_{i,j}$, denoted by $y_{i,j-r}, ..., y_{i,j}$, the various types of transitions can be displayed through a matrix of the type shown as follows

$Y_{i,j-r}$	$Y_{i,j-(r-1)}$	...	$Y_{i,j-1}$	$Y_{ij} = j'$	
				0	1
0	0	...	0	0	1
.	.	...	.	.	
0	0	...	0	0	1
.	.	.	.	.	.
.					
.					
$s_{m,i,j-r}$	$s_{m,i,j-(r-1)}$	...	$s_{m,i,j}$ =0		$s_{m,i,j}$ =1
.	.	...	.	.	
.					
.					
1	1	...	1	0	1

In this matrix, m=1,2,….., 2^r. Let us consider a Markov model of order 3 (r=3). Then m=1,2,…, 2^3 (=8). The following table shows the different types of transitions:

m	States at times Y_{ij-3} Y_{ij-2} Y_{ij-1}	Y_{ij} 0	1
1	0 0 0	π_{0000}	π_{0001}
2	0 0 1	π_{0010}	π_{0011}
3	0 1 0	π_{0100}	π_{0101}
4	0 1 1	π_{0110}	π_{0111}
5	1 0 0	π_{1000}	π_{1001}

m	States at times Y_{ij-3}	Y_{ij-2}	Y_{ij-1}	Y_{ij} 0	1
6	1	0	1	π_{1010}	π_{1011}
7	1	1	0	π_{1100}	π_{1101}
8	1	1	1	π_{1110}	π_{1111}

The transition probability for the transition type $Y_{i,j-r} = s_{m,i,j-r}$,

$Y_{i,j-(r-1)} = s_{m,i,j-(r-1)}, \ldots\ldots, Y_{i,j-1} = s_{m,i,j-1}, Y_{ij} = s_{m,i,j} = j'$ is

$$\pi_{mij}(X) = e^{g_m(X_{ij})} / (1 + e^{g_m(X_{ij})}). \tag{6.2}$$

We can further simplify (6.2) as

$\pi_{mij}(X) = e^{g_m(X_{ij})} / (1 + e^{g_m(X_{ij})})$, and $1 - \pi_{mij}(X) = 1/(1 + e^{g_m(X_{ij})})$,

where $g_m(X_{ij}) = \beta'_m X_{ij}$, $m = 1, 2, \ldots\ldots, 2^r$.

6.3 Likelihood Function

The likelihood function, as a generalization of (5.2), can be expressed as

$$L = \prod_{m=1}^{2^r} \prod_{i=1}^{n} \prod_{j=1}^{n_i} \left[\left\{ \pi_{mij} \right\}^{\delta_{mij}} \left\{ 1 - \pi_{mij} \right\}^{1-\delta_{mij}} \right]. \tag{6.3}$$

From (6.3) the log-likelihood function can be shown as

$$\ln L = \ln L_1 + \ln L_2 + \ldots. + \ln L_m + \ldots\ldots. + \ln L_{2^r}. \tag{6.4}$$

In (6.4),

$$L_m = \prod_{i=1}^{n} \prod_{j=1}^{n_i} \left[\left\{ \pi_{mij} \right\}^{\delta_{mij}} \left\{ 1 - \pi_{mij} \right\}^{1-\delta_{mij}} \right] \tag{6.5}$$

and using (6.2) we express (6.5) as

$$L_m = \prod_{i=1}^{n} \prod_{j=1}^{n_i} \left[\left\{ \frac{1}{1 + e^{\beta'_m X_{ij}}} \right\}^{\delta_{mij}} \left\{ \frac{e^{\beta'_m X_{ij}}}{1 + e^{\beta'_m X_{ij}}} \right\}^{1-\delta_{mij}} \right], \tag{6.6}$$

Using (6.6) the m-th component of the log likelihood function is reduced to

$$\ln L_m = \sum_{i=1}^{n} \sum_{j=1}^{n_i} \left[(1 - \delta_{mij}) \beta'_m X_{ij} - \left(1 + e^{\beta'_m X_{ij}} \right) \right]. \tag{6.7}$$

The first derivatives of (6.7) with respect to parameters are

$$\frac{\partial \ln L_w}{\partial \beta_{mq}} = \sum_{i=1}^{n}\sum_{j=1}^{n_i}\left[X_{ij,q}\left\{(1-\delta_{mij}) - \frac{e^{\beta'_m X_{ij}}}{1+e^{\beta'_m X_{ij}}}\right\}\right], m = 1,2,...,2^r; \text{q=0,1,2,...,p.} \quad (6.8)$$

Solutions of (6.8) give the MLE's of the parameters. The second derivatives for $\ln L_m$ are

$$\frac{\partial^2 \ln L_m}{\partial \beta_{mq} \partial \beta_{ml}} = -\sum_{i=1}^{n}\sum_{j=1}^{n_i}\left[X_{ij,q} X_{ij,l}\left\{\pi_{mij}(1-\pi_{mij})\right\}\right]. \quad (6.9)$$

From (6.9) we get the information matrix.

6.4 Testing for the Significance of Parameters

The vectors of 2^r sets of parameters for the r-th order Markov model can be represented by the following vector:

$$\beta = \left[\beta'_1, \beta'_2, ..., \beta'_{2^r}\right]'$$

where $\beta'_m = \left[\beta_{m0}, \beta_{m1}, ..., \beta_{mp}\right]$, m=1,2,…..,$2^r$.

To test the null hypothesis H_0 about joint vectors of all the regression parameters (excluding intercepts), we can employ the usual likelihood ratio test

$$-2[\ln L(\beta_0) - \ln L(\beta)] \sim \chi^2_{2^r p}. \quad (6.10)$$

To test the significance of the q-th (q=0,1,…,p) parameter of the m-th (m=1,2,…, 2^r) set of parameters, the null hypothesis is $H_0 : \beta_{mq} = 0$ and the corresponding Wald test is

$$W = \hat{\beta}_{mq} / se(\hat{\beta}_{mq}). \quad (6.11)$$

6.5 Applications

We have used the HRS data on mobility of elderly population for the period 1992-2004 to display the fitting of higher order Markov models. The first and second order Markov models are already presented in the previous chapters. This chapter extends the application to the third and fourth orders. The study was initiated in 1992 on the elderly population of the USA and subsequently followed-up every two years. A total of 12,652 respondents were included in this cohort. In all the waves, mobility index for the elderly population was constructed on the basis of five tasks including walking several blocks, walking one block, walking across the room, climbing several flights of stairs, climbing one flight of stairs. For no difficulty in the five tasks the index is 0 and for at least one difficulty the index is 1 or higher. To demonstrate a two-state Markov chain model for the third and fourth orders, we have considered 0 for no difficulty and 1 for difficulty. Table 6.1 displays the pooled number of transitions during two consecutive follow-ups during the period 1992-2004.

Table 6.1 shows the transition counts and transition probabilities for a two-state third order Markov chain. As we observed before for first and second order Markov chains, the transition probabilities of staying in the same state are very high. The transition probabilities for staying in four consecutive follow-ups in the state of no difficulty (0) is 0.865 and in the state of difficulty (1) is 0.908. Reversal to the same states are shown in transitions 0-0-1-0, 1-1-0-1, 0-1-1-0 and 1-0-0-1 with corresponding transition probabilities 0.484, 0.539, 0.262, and 0.340 respectively. It is noteworthy that the reverse transition to difficulty in mobility is relatively higher as compared to that of the reverse transition to no difficulty. Similarly, the repeated transitions to difficulty in mobility (1) is shown in the transition type 0-1-0-1 and the corresponding transition probability is 0.434, while the transition probability of repeated transition to no difficulty (0), which is shown in transition type 1-0-1-0, is 0.303. In other words, the repeated transition to difficulty is higher as compared to that of the repeated transition to no difficulty.

Table 6.1. Transition Counts and Transition Probabilities for Third Order Two State Mobility Index

States			Transition Count		Transition Probability		Total Count
			0	1^+	0	1^+	
0	0	0	10710	1677	0.865	0.135	12387
0	0	1^+	1024	1090	0.484	0.516	2114
0	1^+	0	792	608	0.566	0.434	1400
0	1^+	1^+	495	1391	0.262	0.738	1886
1^+	0	0	768	395	0.660	0.340	1163
1^+	0	1^+	308	709	0.303	0.697	1017
1^+	1^+	0	432	506	0.461	0.539	938
1^+	1^+	1^+	521	5171	0.092	0.908	5692

Now let us consider a test procedure for testing the null hypothesis $H_0 : \pi_{mj'} = \pi^0_{mj'}$ and the test statistic is

$$\chi^2 = \sum_{m=1}^{2^3} \sum_{j'=0}^{1} \frac{n_{m.}(\hat{\pi}_{mj'} - \pi^0_{mj'})^2}{\pi^0_{mj'}}$$

which is χ^2 with 2.2.2(2-1)-d degrees of freedom (number of states=2) under H_0 where d is the number of zero $\pi^0_{mj'}$ s. Here, m represents each of the 8 three prior states as shown in Table 6.1, for instance, 0-0-0, 0-0-1, 0-1-0, 0-1-1, 1-0-0, 1-0-1, 1-1-0, and 1-1-1; j denotes outcome of the jth (more precisely, current) follow-up or, in other words, $Y_{ij} = j'$, $j' = 1$, if the event occurs to the ith individual at the jth (current) follow-up, $j' = 0$, otherwise.

Similarly, the likelihood ratio test shows

$$-2\ln\Lambda = 2\sum_{m=1}^{2^3}\sum_{j'=0}^{1} n_{mj'}\ln\frac{n_{mj'}}{n_{m.}\pi^{0}_{mj'}}.$$

This is χ^2 with 2.2.2(2-1) degrees of freedom under H_0.

Table 6.2 summarizes the results based on the chi square and likelihood ratio tests for $H_0 : \pi_{mj'} = 0.5$. The results indicate that the third order transition probabilities are not 0.5.

Table 6.2. Test for the Third Order Two State Mobility Index

Test-statistics	Value	df	p-value
Chi-square	11121.403	8	0.000
LRT	12515.531	8	0.000

Now let us test for the stationarity of transition probabilities for the third order. We have used two tests here: (i) test for stationarity on the basis of pooled transition data, and (ii) test for stationarity based on consecutive follow-up data.

(v) Test for Stationarity Based on Pooled Transition Data

The null hypothesis is $H_0 : \pi^{t}_{mj'} = \pi_{mj'}$ and the likelihood ratio chi square is

$$-2\ln\Lambda = 2\sum_{t=1}^{T-3}\sum_{m=1}^{2^3}\sum_{j'=0}^{1} n^{t}_{mj'}\ln\frac{n^{t}_{mj'}}{n^{t-1}_{m.}\pi_{mj'}}.$$

where $\pi_{mj'}$ represents the transition probabilities for the pooled time.

This is χ^2 with (T-3)s.s.s.(s-1) degrees of freedom (T=7,s=2) under H_0. Using the pooled estimate of transition probabilities as the null hypothesis values, we observe from Table 6.3 that the chi square is 61.287. The degrees of freedom is 32 and the corresponding p-value is less than 0.001. Hence, the null hypothesis of stationarity for the third order transition probability can be rejected for the given data.

(vi) Test for Stationarity Based on Consecutive Follow-up Data

This test procedure allows us to understand the process of stationarity better. Under stationarity, the consecutive transition probabilities with respect to time should remain constant. Any deviation from the constant transition probabilities at consecutive follow-up times would shed doubt regarding the stationarity of the underlying stationary process.

In this case, the null hypothesis is $H_0 : \pi^{t}_{mj'} = \pi^{t+1}_{mj'}$ where

$\pi^{t}_{mj'}$ = transition probability for transition type mj′ for order 3 at time t and

$\pi^{t+1}_{mj'}$ = transition probability for transition type mj′ for order 3 at time t+1

The likelihood ratio test statistic is

$$-2\ln\Lambda = 2\sum_{t=1}^{T-4}\sum_{j'=0}^{1} n^{t}_{mj'} \ln \frac{n^{t}_{mj'}}{n^{t-1}_{m.}\,\pi^{t+1}_{mj'}}.$$

The degrees of freedom for the χ^2 based on the likelihood ratio test is (T-4)s.s.s(s-1) under H_0 for T=7 and s=2. It is evident from Table 6.3 that the value of chi square is 85.813 mostly attributable to the fourth and the fifth follow-up times. The degrees of freedom is 24 and the p-value is less than 0.001. In other words, we may reject the null hypothesis of stationarity for the given data on mobility index.

Table 6.3. Stationarity Test for the Third Order Two State Mobility Index

Time	D.F.	Comparison Between Consecutive time points		Comparison with polled transition matrix	
		Chi-square	p-value	Chi-square	p-value
4	8	29.932	0.000	21.613	0.006
5	8	42.156	0.000	11.603	0.771
6	8	13.725	0.953	11.973	0.980
7	8			16.098	0.991
Overall χ^2	24,32	85.813	0.000	61.287	0.001

Table 6.4 displays the results of fitting the models for 8 transition types. Due to branching of different transition types, some of the models do not show any significant results. We have used only a few variables here in order to keep the tables simple for understanding. With inclusion of a few more potential risk factors, a more meaningful set of results could be displayed. In the third order models, age, gender or both show significant association with transitions of different types except 0-0-1-0 and 1-1-1-0.

Table 6.5 shows the transition count and transition probabilities for the fourth order Markov model. There are 16 transition types in the four prior transitions which are: 0-0-0-0, 0-0-0-1, 0-0-1-0, 0-0-1-1, 0-1-0-0, 0-1-0-1, 0-1-1-0, 0-1-1-1, 1-0-0-0, 1-0-0-1, 1-0-1-0, 1-0-1-1, 1-1-0-0, 1-1-0-1, 1-1-1-0, and 1-1-1-1.

Table 6.4. Estimates of Parameters of the Logistic Regression Models for the Third Order Two State Mobility Index

Variables	Coeff.	Std. err.	t-value	p-value	95% C.I. LL	95% C.I. UL
FIRST ORDER						
TRANSITION TYPE: 0→0→0→1						
Constant	-3.544	0.427	-8.291	0.000	-4.382	-2.706
Age	0.030	0.007	4.385	0.000	0.017	0.044
Gender	-0.285	0.053	-5.406	0.000	-0.388	-0.182
White	0.044	0.157	0.280	0.779	-0.264	0.352
Black	0.358	0.169	2.117	0.034	0.027	0.689
TRANSITION TYPE: 0→0→1→0						
Constant	-1.199	0.676	-1.775	0.076	-2.524	0.125
Age	0.012	0.011	1.081	0.280	-0.010	0.033
Gender	0.066	0.088	0.754	0.451	-0.106	0.239
White	0.441	0.259	1.704	0.088	-0.066	0.948
Black	0.420	0.275	1.529	0.126	-0.118	0.959
TRANSITION TYPE: 0→1 →0→1						
Constant	-1.158	0.859	-1.349	0.177	-2.841	0.525
Age	0.014	0.014	1.013	0.311	-0.013	0.042
Gender	-0.234	0.110	-2.123	0.034	-0.450	-0.018
White	0.195	0.286	0.680	0.497	-0.367	0.756
Black	0.031	0.311	0.098	0.922	-0.579	0.640
TRANSITION TYPE: 0→ 1→1→0						
Constant	-2.011	0.833	-2.414	0.016	-3.644	-0.378
Age	0.017	0.014	1.275	0.202	-0.009	0.044
Gender	0.050	0.107	0.468	0.640	-0.160	0.261
White	-0.026	0.290	-0.089	0.929	-0.595	0.543
Black	-0.093	0.309	-0.301	0.763	-0.700	0.513
Transition Type: 1→0→0→1						
Constant	-3.781	1.000	-3.779	0.000	-5.742	-1.820
Age	0.060	0.016	3.688	0.000	0.028	0.092
Gender	-0.202	0.128	-1.577	0.115	-0.453	0.049
White	-0.390	0.317	-1.229	0.219	-1.011	0.232
Black	-0.101	0.341	-0.295	0.768	-0.769	0.568
Transition Type: 1→0→1→0						
Constant	0.146	1.081	0.135	0.893	-1.974	2.266
Age	-0.005	0.018	-0.289	0.773	-0.040	0.030
Gender	0.125	0.143	0.874	0.382	-0.155	0.405
White	-0.762	0.334	-2.280	0.023	-1.417	-0.107
Black	-0.723	0.365	-1.983	0.047	-1.438	-0.008
Transition Type: 1→1→0→1						
Constant	0.234	1.127	0.208	0.836	-1.976	2.444
Age	0.009	0.018	0.501	0.616	-0.026	0.045
Gender	-0.284	0.137	-2.072	0.038	-0.553	-0.015
White	-0.555	0.380	-1.459	0.145	-1.300	0.191
Black	-0.362	0.405	-0.893	0.372	-1.155	0.432

Table 6.4. Continued

Variables	Coeff.	Std. err.	t-value	p-value	95% C.I.	
					LL	UL
		FIRST ORDER				
Transition Type: 1→1→1→0						
Constant	-1.065	0.757	-1.407	0.159	-2.549	0.418
Age	-0.020	0.012	-1.640	0.101	-0.045	0.004
Gender	0.100	0.099	1.010	0.313	-0.094	0.293
White	-0.089	0.236	-0.375	0.708	-0.551	0.374
Black	-0.037	0.250	-0.147	0.883	-0.527	0.454
Model χ^2 (df=10)			11198.47 (0.000)			
LRT (df=10)			12632.66 (0.000)			

Table 6.5. Transition Counts and Transition Probabilities for the Fourth Order Two State Mobility Index

States				Transition Count		Transition Probability		Total Count
				0	1	0	1	
0	0	0	0	6867	1003	0.873	0.127	7870
0	0	0	1	591	570	0.509	0.491	1161
0	0	1	0	425	296	0.589	0.411	721
0	0	1	1	213	507	0.296	0.704	720
0	1	0	0	373	199	0.652	0.348	572
0	1	0	1	141	276	0.338	0.662	417
0	1	1	0	167	183	0.477	0.523	350
0	1	1	1	176	813	0.178	0.822	989
1	0	0	0	389	167	0.700	0.300	556
1	0	0	1	102	148	0.408	0.592	250
1	0	1	0	93	122	0.433	0.567	215
1	0	1	1	107	345	0.237	0.763	452
1	1	0	0	182	128	0.587	0.413	310
1	1	0	1	85	255	0.250	0.750	340
1	1	1	0	148	218	0.404	0.596	366
1	1	1	1	214	3102	0.065	0.935	3316

Now let us consider a test procedure for testing the null hypothesis $H_0 : \pi_{mj'} = \pi^0_{mj'}$ and the test statistic is

$$\chi^2 = \sum_{m=1}^{2^4} \sum_{j'=0}^{1} \frac{n_{m.}(\hat{\pi}_{mj'} - \pi^0_{mj'})^2}{\pi^0_{mj'}}$$

which is χ^2 with 2.2.2.2(2-1)-d degrees of freedom (s=number of states=2) under H_0 where d is the number of zeros in π^0. Here, m represents each of the 16 four prior states as shown in Table 6.5; j′ denotes outcome of the j-th (more precisely, current) follow-up or, in other words, $Y_{ij} = j'$, $j' = 1$, if the event occurs to the i-th individual at the j-th (current) follow-up, $j' = 0$, otherwise.

Similarly, the likelihood ratio test shows

$$-2\ln\Lambda = 2\sum_{m=1}^{2^4}\sum_{j'=0}^{1} n_{mj'} \ln\frac{n_{mj'}}{n_{m.}\pi_{mj'}^{0}}.$$

This is χ^2 with 2.2.2.2(2-1) degrees of freedom under H_0.

Table 6.6 summarizes the results based on the chi square and likelihood ratio tests for $H_0 : \pi_{mj'} = 0.5$. The results indicate that the third order transition probabilities are not 0.5.

Table 6.6. Test for the Fourth Order Two State Mobility Index

Test-statistics	Value	D.F.	p-value
Chi-square	7869.84366	16	0.000000
LRT	8953.75616	16	0.000000

Now let us test for the stationarity of transition probabilities for the third order. We have used two tests here: (i) test for stationarity on the basis of pooled transition data, and (ii) test for stationarity based on consecutive follow-up data.

(i) Test for Stationarity Based on Pooled Transition Data

The null hypothesis is $H_0 : \pi_{mj'}^{t} = \pi_{mj'}$ and the likelihood ratio chi square is

$$-2\ln\Lambda = 2\sum_{t=1}^{T-4}\sum_{m=1}^{2^4}\sum_{j'=0}^{1} n_{mj'}^{t} \ln\frac{n_{mj'}^{t}}{n_{m.}^{t-1}\pi_{mj'}}.$$

where $\pi_{mj'}$ represents the transition probabilities for the pooled time.

This is χ^2 with (T-4)s.s.s.s.(s-1) degrees of freedom (s=2) under H_0. Using the pooled estimate of transition probabilities as the null hypothesis values, we observe from Table 6.7 that the chi square is 44.850. The degrees of freedom is 48 and the corresponding p-value is 0.603. Hence, the null hypothesis of stationarity for the fourth order transition probability can be accepted for the given data.

(ii) Test for Stationarity Based on Consecutive Follow-up Data

This test procedure allows us to understand the process of stationarity better. Under stationarity, the consecutive transition probabilities with respect to time should remain constant. Any deviation from the constant transition probabilities at consecutive follow-up times would shed doubt regarding the stationarity of the underlying stationary process.

In this case, the null hypothesis is $H_0 : \pi^{t}_{mj'} = \pi^{t+1}_{mj'}$ where

$\pi^{t}_{mj'}$ = transition probability for transition type mj′ for order 3 at time t and

$\pi^{t+1}_{mj'}$ = transition probability for transition type mj′ for order 3 at time t+1

The likelihood ratio test statistic is

$$-2\ln\Lambda = 2\sum_{t=1}^{T-5}\sum_{j'=0}^{1} n^{t}_{mj'}\ln\frac{n^{t}_{mj'}}{n^{t-1}_{m.}\,\pi^{t+1}_{mj'}}.$$

The degrees of freedom for the χ^2 based on the likelihood ratio test is (T-5)s.s.s.s(s-1) under H_0 (T=7, s=2). It is evident from Table 6.7 that the value of chi square is 93.655.The degrees of freedom is 32 and the p-value is less than 0.001. In other words, we may reject the null hypothesis of stationarity for the given data on mobility index. This result is different from that of the test based on the pooled data which is indicative of the fact that there is quite fluctuation in the subsequent follow-up movements even for the fourth order transitions. However, these fluctuations are eliminated in the pooled data.

Table 6.7. Stationary Test for the Fourth Order Two State Mobility Index

Time	D.F.	Comparison between Consecutive Time Points		Comparison with Pooled Transition Matrix	
		Chi-square	p-value	Chi-square	p-value
5	16	57.301622	0.000001	17.907209	0.329359
6	16	36.354036	0.002583	13.422022	0.641687
7	16			13.520985	0.634352
Overall χ^2	32,48	93.655658	0.000000	44.850216	0.602684

Table 6.8 shows the estimates of the parameters for the 16 models generated from the fourth order Markov model. To keep the results simple, we have used only two covariates, age and gender. The interpretations are similar to that of the results presented in Chapter 4 and 5, so we are not illustrating the results in this chapter. It is noteworthy that if we include a few more important covariates, the results would be more meaningful and the fit would be much better for different types of transitions.

Table 6.8. Estimates of the Parameters of the Logistic Regression Models for the Fourth Order Two State Mobility Index

Variables	Coeff.	Std. error	t-value	p-value	95% C. I.	
					LL	UL
0 → 0 → 0→ 0→ 1						
Constant	-3.625	0.552	-6.573	0.000	-4.706	-2.544
Age	0.032	0.010	3.349	0.001	0.013	0.051
Gender	-0.249	0.068	-3.671	0.000	-0.382	-0.116
0 → 0 → 0→ 1→ 0						
Constant	-0.861	0.925	-0.931	0.352	-2.674	0.952
Age	0.017	0.016	1.028	0.304	-0.015	0.048
Gender	-0.108	0.118	-0.919	0.358	-0.339	0.122
0 → 0 → 1→ 0→ 1						
Constant	-0.238	1.213	-0.196	0.844	-2.616	2.139
Age	0.000	0.021	0.004	0.997	-0.041	0.041
Gender	-0.280	0.153	-1.838	0.066	-0.579	0.019
0 → 0 → 1→ 1→ 0						
Constant	-3.097	1.280	-2.420	0.016	-5.605	-0.589
Age	0.042	0.022	1.853	0.064	-0.002	0.085
Gender	-0.308	0.167	-1.847	0.065	-0.635	0.019
0 → 1 → 0→ 0→ 1						
Constant	-3.645	1.399	-2.605	0.009	-6.388	-0.903
Age	0.054	0.024	2.214	0.027	0.006	0.101
Gender	-0.167	0.179	-0.934	0.350	-0.519	0.184
0 → 1 → 0→ 1→ 0						
Constant	-0.788	1.731	-0.455	0.649	-4.182	2.606
Age	0.001	0.030	0.042	0.967	-0.058	0.060
Gender	0.117	0.213	0.549	0.583	-0.300	0.533
0 → 1 → 1→ 0→ 1						
Constant	-2.114	1.755	-1.204	0.229	-5.555	1.327
Age	0.041	0.031	1.335	0.182	-0.019	0.101
Gender	-0.335	0.221	-1.513	0.130	-0.769	0.099
0 → 1 → 1→ 1→ 0						
Constant	-2.289	1.367	-1.674	0.094	-4.969	0.391
Age	0.013	0.024	0.538	0.591	-0.034	0.060
Gender	0.071	0.173	0.414	0.679	-0.267	0.410
1 → 0 → 0→ 0→ 1						
Constant	-3.533	1.601	-2.207	0.027	-6.671	-0.396
Age	0.046	0.028	1.656	0.098	-0.008	0.101
Gender	0.092	0.187	0.493	0.622	-0.275	0.459
1 → 0 → 0→ 1→ 0						
Constant	1.761	2.034	0.865	0.387	-2.227	5.748
Age	-0.037	0.035	-1.062	0.288	-0.105	0.031
Gender	0.032	0.266	0.120	0.905	-0.489	0.553
1 → 0 → 1→ 0→ 1						
Constant	1.275	2.235	0.570	0.568	-3.106	5.655
Age	-0.016	0.039	-0.407	0.684	-0.092	0.061
Gender	-0.258	0.287	-0.899	0.369	-0.819	0.304
1 → 0 → 1→ 1→ 0						
Constant	0.771	1.777	0.434	0.664	-2.712	4.255

Table 6.8 (Cont'd)

Age	-0.034	0.031	-1.091	0.275	-0.094	0.027
Gender	-0.025	0.230	-0.107	0.915	-0.476	0.427
1 → 1 → 0 → 0 → 1						
Constant	-2.958	1.981	-1.493	0.135	-6.840	0.925
Age	0.048	0.034	1.412	0.158	-0.019	0.115
Gender	-0.460	0.241	-1.908	0.056	-0.933	0.013
1 → 1 → 0 → 1 → 0						
Constant	-1.145	2.005	-0.571	0.568	-5.076	2.786
Age	0.001	0.035	0.020	0.984	-0.067	0.069
Gender	0.019	0.272	0.069	0.945	-0.515	0.553
1 → 1 → 1 → 0 → 1						
Constant	0.938	1.890	0.496	0.620	-2.767	4.643
Age	-0.008	0.033	-0.256	0.798	-0.073	0.056
Gender	-0.214	0.228	-0.941	0.347	-0.661	0.232
1 → 1 → 1 → 1 → 0						
Constant	-1.546	1.169	-1.322	0.186	-3.839	0.746
Age	-0.019	0.020	-0.942	0.346	-0.059	0.021
Gender	-0.091	0.158	-0.578	0.563	-0.401	0.218
Global Chi-square			7913.90	0.000		
LRT			9015.46	0.000		

6.6 Summary

This chapter is a generalization of Chapters 4 and 5. The higher order (order greater than 2) Markov models with covariate dependence are studied in this chapter. The inferential procedures are also generalized and illustration is provided for mobility index transitions among elderly population of the USA. The results are displayed for the third and fourth orders of the underlying Markov models.

Chapter 7

MULTISTATE FIRST ORDER MARKOV MODEL WITH COVARIATE DEPENDENCE

7.1 INTRODUCTION

In the previous chapters, we have considered only binary outcomes. However, in real life situations, we may have to consider multistate models with more than two outcomes. Some examples of multiple outcomes are: (i) disease status of an individual (normal, mild, severe), (ii) rainfall amount (normal, moderate, and heavy), (iii) transaction in share market (normal, moderate, and heavy), etc. These examples represent states of three-state Markov models which can be generalized further by increasing the number of states. In a k-state Markov model, the underlying transition probabilities are represented by a $k \times k$ matrix. The number of models we need to consider is (k-1) from each row of transition matrix.

7.2 MULTISTATE MARKOV MODEL

In the previous chapters, we have considered only binary outcomes. In this section, we extend the theory to multiple outcomes for a single stationary process $(Y_{i1}, Y_{i2}, ..., Y_{ij})$ representing the past and present responses for subject i (i= 1,.2,…,n) at follow-up j (j=1,2,…, n_i). Here, as defined earlier, Y_{ij} is the response at time t_{ij}.

The multiple outcomes are defined by Y_{ij} =s, s=0,1,2,…,k-1 if an event of level s occurs for the i-th subject at the j-th follow-up where Y_{ij} =0 indicates that no event occurs. Then the first order Markov model can be expressed as

$$P(Y_{ij} \left| H_{ij} \right.) = P(Y_{ij} \left| Y_{ij-1} \right.),$$

and the corresponding k x k transition probability matrix is given by

$$\pi = \begin{bmatrix} \pi_{00} & \cdots & \pi_{0,k-1} \\ \pi_{10} & \cdots & \pi_{1,k-1} \\ . & & \\ . & & \\ . & & \\ \pi_{k-1,0} & \cdots & \pi_{k-1,k-1} \end{bmatrix}$$

where the row probabilities add to 1 . Here, 0,1,…,k-1 are the k possible outcomes of the variable, Y. The probability of a transition from u (u=0,…,k-1) at time $t_{i,j-1}$ to s (s=0,…,k-1) at time t_{ij} is $\pi_{us} = P(Y_{ij} = s \mid Y_{i,j-1} = u)$. For any k,

$$\sum_{s=0}^{k-1} \pi_{us} = 1, \text{ u=0,..., k-1.}$$

Let us consider a vector of p-variables for j-th follow-up and the corresponding vector of parameters, as follows:

$X_{ij} = \left[1, \mathrm{X}_{\mathrm{ij1}}, \text{ ...., } \mathrm{X}_{\mathrm{ij,p}}\right]$ = vector of covariates for the i-th person at j-th follow-up;

$\beta'_{us} = \left[\beta_{us0}, \beta_{\mathrm{us1}}, \text{ ...., } \beta_{\mathrm{usp}}\right]$ = vector of parameters for the transition from u to s.

The probabilities of transition from state u to state s can be expressed in terms of conditional probabilities (Hosmer and Lemeshow, 1989) as follows:

$$\pi_{usij} = P(Y_{ij} = s \mid Y_{i,j-1} = u, X_{ij}) = \frac{e^{g_{us}(X_{ij})}}{\sum_{s=0}^{k-1} e^{g_{us}(X_{ij})}}, \text{ u=0,...,k-1} \tag{7.1}$$

where

$$g_{us}(X_{ij}) = \begin{cases} 0, & \text{if s} = 0 \\ \ln\left[\dfrac{\pi_{\mathrm{us}}(Y_{ij} = s \mid Y_{i,j-1} = u, X_{ij})}{\pi_{\mathrm{us}}(Y_{ij} = 0 \mid Y_{i,j-1} = u, X_{ij})}\right], & \text{if s} = 1,\ldots,\text{k-1} \end{cases}$$

and

$$g_{us}(X_{ij}) = \beta_{us0} + \beta_{us1} X_{ij1} + \ldots + \beta_{usp} X_{ijp}\,.$$

7.3 Likelihood Function

Then using (7.1) the likelihood function for n individuals with i-th individual having n_i (i=1,2,….n) follow-ups can be expressed as

$$L = \prod_{i=1}^{n} \prod_{j=1}^{n_i} \prod_{u=0}^{k-1} \prod_{s=0}^{k-1} \left[\left\{ \pi_{usij} \right\}^{\delta_{usij}} \right] \tag{7.2}$$

where δ_{usij}=1 if a transition type $u \rightarrow s$ is observed during j-th follow-up for the i-th individual, δ_{usij}=0, otherwise, u,s=0,…,k-1. From (7.2) the log likelihood function is given by

$$\ln L = \sum_{u=0}^{k-1} \ln \mathrm{L_u} \ ,$$

where L_u corresponds to the u-th component of the likelihood function and

$$\ln L_u = \sum_{i=1}^{n} \sum_{j=1}^{n_i} [\sum_{s=0}^{k-1} \delta_{usij} g_{us}(X_{ij}) - \ln\left(\sum_{s=0}^{k-1} e^{g_{us}(X_{ij})} \right)]. \tag{7.3}$$

Substituting (7.3) in $\ln L$ and differentiating with respect to the parameters we obtain the following equations

$$\frac{\partial \ln L_u}{\partial \beta_{usq}} = \sum_{i=1}^{n} \sum_{j=1}^{n_i} X_{qij}(\delta_{usij} - \pi_{usij}) = 0 \ , \quad \text{q=0,1,2,….,p; u=0,…,k-1} \tag{7.4}$$

and solving (7.4) we obtain the likelihood estimates for the 2(k-1)(p+1) parameters. The second derivatives of the log likelihood are given by

$$\frac{\partial^2 \ln L_u}{\partial \beta_{usq} \partial \beta_{usq'}} = -\sum_{i=1}^{n} \sum_{j=1}^{n_i} X_{q'ij} X_{qij} \pi_{usij} (1 - \pi_{usij}) \ , \tag{7.5}$$

q, q'=0,1,2,….,p; s=0,…,k-1; u=0,…,k-1,

and

$$\frac{\partial^2 \ln L_u}{\partial \beta_{usq} \partial \beta_{us'q'}} = -\sum_{i=1}^{n} \sum_{j=1}^{n_i} X_{q'ij} X_{qij} \pi_{usij} \pi_{us'ij} \ , \tag{7.6}$$

q, q'=0,1,2,….,p; s, s'=0,…,k-1; u=0,…,k-1.

The observed information matrix can be obtained from (7.5) and (7.6).

7.4 Testing for the Significance of Parameters

The vectors of $k(k-1)$ sets of parameters for the k-state Markov model of the first order can be represented by the following vector:

$$\beta = \left[\beta_1', \beta_2', \ldots\ldots\ldots, \beta_{k(k-1)}' \right]'$$

where $\beta_v' = \left\lfloor \beta_{v0}, \beta_{v1}, \ldots\ldots, \beta_{vp} \right\rfloor$, v=1,2,…..,$k(k-1)$..

To test the null hypothesis H_0 about joint vanishing of all the regression parameters (excluding intercepts), we can employ the usual likelihood ratio test

$$-2[\ln L(\beta_0) - \ln L(\beta)] \sim \chi^2_{k(k-1)p}.$$

To test the significance of the q-th (q=0,1,…,p) component of the v-th (v=1,2,…,k(k-1)) set of parameters, the null hypothesis is $H_0 : \beta_{vq} = 0$ and the corresponding Wald test statistic is $W = \hat{\beta}_{vq} / se(\hat{\beta}_{vq})$.

7.5 Applications

In this chapter, we have used the same HRS data on mobility of elderly population for the period 1992-2004 to display the fitting of multi-state Markov models. We have defined the outcome variable based on the difficulty in mobility of elderly population. In all the waves, mobility index for the elderly population was constructed on the basis of five tasks including walking several blocks, walking one block, walking across the room, climbing several flights of stairs, climbing one flight of stairs. For the application of a three state model, we have considered 0= no difficulty, 1= difficulty in one of the five tasks, 2= difficulty in performing two or more tasks. Similarly, for a four state model, we have considered, 0= no difficulty, 1= difficulty in one of the five tasks, 2= difficulty in performing two tasks, and 3= difficulty in performing 3 or more tasks. Table 7.1 displays the pooled number of transitions during two consecutive follow-ups in the period 1992-2004 for a three state Markov model. Table 7.1 shows that the probability of remaining in the same state in consecutive follow-ups i.e. 0-0, 1-1 and 2-2, are the largest, 0.800, 0.388 and 0.713, respectively. The transition probabilities from 0 to 1 and from 1 to 2 are 0.136 and 0.257 respectively, in consecutive follow-ups. In a more extreme situation, transition probability from 0 to 2 is 0.064.

Table 7.1. Transition Count and Transition Probability Matrix for the First Order Three State Mobility Index

Mobility Index	Transition Count			Transition Probability			
	0	1	2	0	1	2	Total
0	22461	3824	1797	0.800	0.136	0.064	28082
1	2789	3050	2018	0.355	0.388	0.257	7857
2	944	1496	6072	0.111	0.176	0.713	8512

The test for specified values of the transition probabilities as shown in Table 7.1, we can use the chi square test as described below.

The null hypothesis is $H_0 : \pi_{us} = \pi_{us}^0$ and the test statistic for a three state Markov model is

$$\chi^2 = \sum_{u=0}^{3-1} \sum_{s=0}^{3-1} \frac{n_{u.}(\hat{\pi}_{us} - \pi_{us}^0)^2}{\pi_{us}^0}$$

which is χ^2 with 3(3-1)-d degrees of freedom under H_0 where d is the number of zeros in π^0. If we set the null hypothesis value of π_{us}^0=1/3, then the chi square value is 33525.547 (p-value=0.000). Hence we may reject the null hypothesis that π_{us}^0=1/3.

Similarly, the likelihood ratio test for a three state model is

$$-2\ln\Lambda = 2\sum_{u=0}^{3-1} \sum_{s=0}^{3-1} n_{us} \ln \frac{n_{us}}{n_{u.}\pi_{us}^0}.$$

This is χ^2 with 3(3-1) degrees of freedom under H_0. Table 7.2 shows for π_{us}^0=1/3 that the chi-square value is 32015.062, which is quite large. The null hypothesis may be rejected (p-value=0.000).

Table 7.2. Test for Inference on the First Order Three State Mobility Index

Test-statistics	Value	D.F.	p-value
Chi-square	33525.5476	6	0.000000
LRT	32015.0622	6	0.000000

We can use the test for stationarity for the three state Markov model, like the way we have presented in other chapters. The first method is based on the pooled transition data and the second one is based on the transition probabilities in successive follow-ups. Two test procedures are summarized below and the results are displayed in Table 7.3:

(vii) Test for Stationarity Based on Pooled Transition Data

The null hypothesis is $H_0 : \pi_{us}^t = \pi_{us}$ and the likelihood ratio chi square is

$$-2\ln\Lambda = 2\sum_{t=1}^{T-1} \sum_{u=0}^{3-1} \sum_{s=0}^{3-1} n_{us}^t \ln \frac{n_{us}^t}{n_{u.}^{t-1}\pi_{us}}.$$

where π_{us} represents the transition probabilities for the pooled time, T=7.

This is χ^2 with (7-1)3(3-1)=36 degrees of freedom under H_0. Table 7.3 shows that the value of chi square for the data under study is 100.842 with and the corresponding p value is less than 0.001. Hence, the null hypothesis of stationarity can be rejected for the three state Markov model for the given data.

(viii) Test for Stationarity Based on Consecutive Follow-up Data

The null hypothesis is $H_0 : \pi_{us}^{t} = \pi_{us}^{t+1}$ where

π_{us}^{t} = transition probability from u to s for order 1 at time t and

π_{us}^{t+1} = transition probability from u to s for for order 1 at time t+1

then the likelihood ratio test statistic is

$$-2\ln\Lambda = 2\sum_{t=1}^{T-1}\sum_{u=0}^{3-1}\sum_{s=0}^{3-1} n_{us}^{t} \ln\frac{n_{us}^{t}}{n_{u.}^{t-1}\pi_{us}^{t+1}}.$$

This is χ^2 with (7-2)3(3-1)=30 degrees of freedom under H_0. The chi-square based on the consecutive follow-up data is 101.409. The corresponding p value for chi square with 30 degrees of freedom is less than 0.001. The null hypothesis of stationarity for the three state Markov model may be rejected under this test procedure as well.

Table 7.3. Stationary Test for the First Order Three State Mobility Index

Time	D.F.	Comparison between Consecutive Time Points		Comparison with Pooled Transition Matrix	
		Chi-square	p-value	Chi-square	p-value
2	6	32.784013	0.000012	19.662893	0.003179
3	6	22.694355	0.000906	13.583227	0.034655
4	6	16.794682	0.010068	24.323891	0.000455
5	6	23.560164	0.000629	3.269261	0.774372
6	6	5.575859	0.472336	14.470554	0.024799
7	6			25.532697	0.000272
Overall χ^2	30,36	101.409073	0.000000	100.842523	0.000000

The covariate dependent three state Markov models are fitted to the mobility index for the elderly population. In total, there are six models fitted, for transition types shown in Table 7.4. It is evident that both for transition types 0-1 and 0-2, age is positively and gender is negatively associated. In addition, marginally (p-value<0.10) White race shows negative and Black race shows positive association as compared to that of other races for the transition type 0-2. For reverse transitions, 1-0 and 2-0, gender shows positive association, and age shows negative association only for 2-0 type of transition. Similarly, age is negatively associated with transition from 2 to 1.

Table 7.4. Estimates of the Parameters of the Logistic Regression Models for the Three State First Order Markov Model for Mobility Index

Variables	Coeff.	Std. error	t-value	p-value	95% C. I.	
					LL	UL
0 → 1						
Constant	-2.283	0.245	-9.337	0.000	-2.763	-1.804
Age	0.015	0.004	3.902	0.000	0.007	0.022
Gender	-0.581	0.035	-16.407	0.000	-0.651	-0.512
White	-0.135	0.095	-1.418	0.156	-0.322	0.052
Black	0.090	0.103	0.872	0.383	-0.112	0.292
0 → 2						
Constant	-3.796	0.341	-11.119	0.000	-4.465	-3.127
Age	0.027	0.005	4.995	0.000	0.016	0.037
Gender	-0.362	0.049	-7.380	0.000	-0.458	-0.266
White	-0.220	0.131	-1.680	0.093	-0.476	0.037
Black	0.263	0.139	1.889	0.059	-0.010	0.536
1 → 0						
Constant	0.318	0.337	0.944	0.345	-0.342	0.979
Age	-0.008	0.005	-1.465	0.143	-0.018	0.003
Gender	0.312	0.049	6.360	0.000	0.216	0.408
White	-0.089	0.131	-0.679	0.497	-0.346	0.168
Black	0.069	0.141	0.487	0.626	-0.207	0.344
1 → 2						
Constant	-0.247	0.368	-0.673	0.501	-0.968	0.473
Age	-0.001	0.006	-0.126	0.900	-0.012	0.010
Gender	0.195	0.054	3.598	0.000	0.089	0.301
White	-0.268	0.138	-1.938	0.053	-0.539	0.003
Black	0.131	0.148	0.888	0.374	-0.158	0.421
2 → 0						
Constant	-0.212	0.499	-0.426	0.670	-1.190	0.765
Age	-0.026	0.008	-3.373	0.001	-0.042	-0.011
Gender	0.408	0.070	5.818	0.000	0.271	0.546
White	-0.209	0.165	-1.264	0.206	-0.533	0.115
Black	-0.247	0.176	-1.405	0.160	-0.593	0.098
2 → 1						
Constant	-0.434	0.412	-1.052	0.293	-1.242	0.375
Age	-0.015	0.006	-2.360	0.018	-0.028	-0.003
Gender	0.016	0.060	0.274	0.784	-0.101	0.134
White	0.003	0.144	0.019	0.985	-0.280	0.285
Black	-0.264	0.154	-1.712	0.087	-0.566	0.038
Model χ^2 (df=30)		17402.33	0.00000			
LRT (df=10)		21692.63	0.00000			

The four state transition counts and transition probabilities are shown in Table 7.5. The transition probabilities of remaining in states 0, 1, 2 and 3 are 0.800, 0.388, 0.290 and 0.637 respectively. We also observe that substantial proportions move from 1 to 0, 2 to 3 and 2 to 1, and 3 to 2.

Table 7.5. Transition Count and Transition Probability Matrix for the First Order Four State Mobility Index

Mobility Index	Transition Count				Transition Probability				
	0	**1**	**2**	**3**	**0**	**1**	**2**	**3**	**Total**
0	22461	3824	1091	706	0.800	0.136	0.039	0.025	28082
1	2789	3050	1231	787	0.355	0.388	0.157	0.100	7857
2	661	1000	1142	1132	0.168	0.254	0.290	0.288	3935
3	283	496	883	2915	0.062	0.108	0.193	0.637	4577

The test for specified values of the transition probabilities as shown in Table 7.5, we can use the chi-square test or the likelihood ratio test as shown below.

The null hypothesis is $H_0 : \pi_{us} = \pi_{us}^0$ and the test statistic for a four state Markov model is

$$\chi^2 = \sum_{u=0}^{4-1} \sum_{s=0}^{4-1} \frac{n_{u.}(\hat{\pi}_{us} - \pi_{us}^0)^2}{\pi_{us}^0}$$

which is χ^2 with 4(4-1)-d degrees of freedom under H_0 where d is the number of zeros in π^0. If we set the null hypothesis value of π_{us}^0=1/4, then the chi square value is 51997.030 (p-value=0.000). Hence we may reject the null hypothesis that π_{us}^0 =1/4.

Similarly, the likelihood ratio test for a three state model is

$$-2\ln\Lambda = 2\sum_{u=0}^{4-1} \sum_{s=0}^{4-1} n_{us} \ln \frac{n_{us}}{n_{u.}\pi_{us}^0}.$$

This is χ^2 with 4(4-1) degrees of freedom under H_0. Table 7.6 shows for π_{us}^0 =1/4 that the chi square value is 45879.032, which is quite large. The null hypothesis may be rejected (p-value=0.000).

Table 7.6. Test for the Transition Probabilities of the First Order Four State Mobility Index

Test-statistics	Value	D.F.	p-value
Chi-square	51997.0304	12	0.000000
LRT	45879.0326	12	0.000000

The test procedure for four state Markov model is illustrated below. The results are presented in Table 7.7

(xvi) Test for Stationarity Based on Pooled Transition Data

The null hypothesis is $H_0 : \pi_{us}^t = \pi_{us}$ and the likelihood ratio chi square is

$$-2\ln\Lambda = 2\sum_{t=1}^{T-1}\sum_{u=0}^{4-1}\sum_{s=0}^{4-1} n_{us}^t \ln\frac{n_{us}^t}{n_{u.}^{t-1}\pi_{us}}.$$

where π_{us} represents the transition probabilities for the pooled time, T=7.

This is χ^2 with (7-1)4(4-1)=72 degrees of freedom under H_0. Table 7.7 shows that the value of chi square for the data under study is 137.207 and the corresponding p value is less than 0.001. Hence, the null hypothesis of stationarity can be rejected for the four state Markov model for the given data.

(x) Test for Stationarity Based on Consecutive Follow-up Data

The null hypothesis is $H_0 : \pi_{us}^t = \pi_{us}^{t+1}$ where

π_{us}^t = transition probability from u to s for order 1 at time t and

π_{us}^{t+1} = transition probability from u to s for for order 1 at time t+1

then the likelihood ratio test statistic is

$$-2\ln\Lambda = 2\sum_{t=1}^{T-2}\sum_{u=0}^{4-1}\sum_{s=0}^{4-1} n_{us}^t \ln\frac{n_{us}^t}{n_{u.}^{t-1}\pi_{us}^{t+1}}.$$

This is χ^2 with (7-2)4(4-1)=60 degrees of freedom under H_0. The chi square based on the consecutive follow-up data is 200.558. The corresponding p value for chi square with 60 degrees of freedom is less than 0.001. The null hypothesis of stationarity for the four state Markov model may be rejected under this test procedure as well.

Table 7.7. Stationary Test for First Order Four State Mobility Index

Time	D.F.	Comparison between Consecutive Time Points		Comparison with Pooled Transition Matrix	
		Chi-square	p-value	Chi-square	p-value
2	12	42.211496	0.000031	23.470464	0.023987
3	12	51.068997	0.000001	19.951368	0.068012
4	12	38.299017	0.000137	35.386619	0.000406
5	12	47.332890	0.000004	6.179042	0.906789
6	12	21.646278	0.041681	24.198323	0.019114
7	12			28.021597	0.005492
Overall χ^2	60,72	200.558678	0.000000	137.207412	0.000006

The estimates of parameters for a four state covariate dependent Markov model are displayed in Table 7.8. In total, we have fitted 12 regression models for independent transitions. We observe that age is positively and gender is negatively associated with transitions from 0 to 1, 0 to 2 and 0 to 3. Gender is positively associated with transition types 1-0 and 1-2. Whites have lower risk of transition from 1 to 3 as compared to that of other races. There is negative association between transition from 2 to 0,1 and 3 and age. In addition, gender is positively associated with transition from 2 to 0. All the transitions from 3, i.e. 3-0, 3-1 and 3-2 are negatively associated with age while gender is positively associated with transition types 3-0 and 3-1.

Table 7.8. Estimates of First Order Four State Markov Model for Mobility Index

Variables	Coeff.	Std. error	t-value	p-value	95% C. I.	
					LL	UL
0 → 1						
Constant	-2.284	0.245	-9.339	0.000	-2.763	-1.804
Age	0.015	0.004	3.903	0.000	0.007	0.022
Gender	-0.581	0.035	-16.406	0.000	-0.651	-0.512
White	-0.135	0.095	-1.416	0.157	-0.322	0.052
Black	0.090	0.103	0.872	0.383	-0.112	0.292
0 → 2						
Constant	-4.788	0.436	-10.989	0.000	-5.643	-3.934
Age	0.034	0.007	4.991	0.000	0.020	0.047
Gender	-0.426	0.062	-6.839	0.000	-0.548	-0.304
White	-0.091	0.174	-0.521	0.602	-0.431	0.250
Black	0.281	0.185	1.516	0.130	-0.082	0.643
0 → 3						
Constant	-3.988	0.527	-7.561	0.000	-5.022	-2.954
Age	0.016	0.008	1.890	0.059	-0.001	0.032
Gender	-0.264	0.076	-3.451	0.001	-0.414	-0.114
White	-0.401	0.191	-2.099	0.036	-0.775	-0.027
Black	0.244	0.203	1.202	0.229	-0.154	0.641
1 → 0						
Constant	0.321	0.337	0.951	0.342	-0.340	0.982
Age	-0.008	0.005	-1.469	0.142	-0.018	0.003
Gender	0.313	0.049	6.386	0.000	0.217	0.409
White	-0.090	0.131	-0.684	0.494	-0.346	0.167
Black	0.069	0.141	0.492	0.623	-0.206	0.345
1 → 2						
Constant	-0.808	0.443	-1.823	0.068	-1.677	0.061
Age	-0.002	0.007	-0.264	0.792	-0.015	0.012
Gender	0.249	0.064	3.857	0.000	0.122	0.375
White	-0.139	0.172	-0.808	0.419	-0.475	0.198
Black	0.201	0.183	1.101	0.271	-0.157	0.559
1 → 3						
Constant	-1.104	0.529	-2.086	0.037	-2.141	-0.067
Age	0.001	0.008	0.117	0.907	-0.015	0.017
Gender	0.110	0.080	1.380	0.167	-0.046	0.266
White	-0.451	0.186	-2.420	0.016	-0.816	-0.086
Black	0.041	0.199	0.205	0.837	-0.349	0.431

Table 7.8. Continued

Variables	Coeff.	Std. error	t-value	p-value	95% C. I.	
					LL	UL
2 → 0						
Constant	0.897	0.604	1.485	0.138	-0.287	2.082
Age	-0.024	0.010	-2.480	0.013	-0.042	-0.005
Gender	0.311	0.087	3.583	0.000	0.141	0.482
White	-0.110	0.199	-0.552	0.581	-0.501	0.281
Black	-0.227	0.216	-1.050	0.294	-0.650	0.196
2 → 1						
Constant	0.585	0.525	1.113	0.266	-0.445	1.615
Age	-0.015	0.008	-1.815	0.069	-0.031	0.001
Gender	-0.115	0.077	-1.488	0.137	-0.267	0.037
White	0.283	0.193	1.466	0.143	-0.095	0.662
Black	0.040	0.206	0.193	0.847	-0.365	0.445
2 → 3						
Constant	0.848	0.503	1.685	0.092	-0.139	1.835
Age	-0.016	0.008	-2.067	0.039	-0.032	-0.001
Gender	-0.051	0.074	-0.689	0.491	-0.196	0.094
White	0.136	0.181	0.752	0.452	-0.219	0.490
Black	0.211	0.191	1.104	0.270	-0.164	0.586
3 → 0						
Constant	0.192	0.912	0.210	0.834	-1.596	1.979
Age	-0.043	0.014	-3.071	0.002	-0.071	-0.016
Gender	0.504	0.125	4.043	0.000	0.259	0.748
White	-0.176	0.321	-0.547	0.584	-0.805	0.454
Black	0.101	0.332	0.305	0.760	-0.549	0.752
3 → 1						
Constant	0.167	0.694	0.241	0.810	-1.192	1.527
Age	-0.027	0.011	-2.493	0.013	-0.048	-0.006
Gender	0.178	0.099	1.801	0.072	-0.016	0.372
White	-0.337	0.220	-1.533	0.125	-0.767	0.094
Black	-0.497	0.235	-2.112	0.035	-0.958	-0.036
3 → 2						
Constant	-0.255	0.553	-0.460	0.645	-1.338	0.829
Age	-0.015	0.009	-1.769	0.077	-0.032	0.002
Gender	0.032	0.079	0.405	0.685	-0.123	0.188
White	-0.017	0.196	-0.087	0.930	-0.401	0.366
Black	-0.096	0.205	-0.465	0.642	-0.498	0.307
Global Chi-square			19854.35 (0.000)			
LRT			27116.55 (0.000)			

7.6 SUMMARY

In the previous chapters, we had shown two state models only. In practice, in many situations we observe more than two states. This chapter generalizes the previous methodology of the covariate dependent first order Markov models of two states to multiple states. The generalization is quite straightforward and the estimation and test procedures are illustrated for 3 and 4 states on the basis of the HRS data on difficulty in movement of the elderly population. The procedures for tests on specified transition probabilities and on stationarity are also generalized in this chapter.

Chapter 8

MULTISTATE MARKOV MODEL OF HIGHER ORDER WITH COVARIATE DEPENDENCE

8.1 INTRODUCTION

In Chapter 7, we introduced the multistate Markov models of first order with covariate dependence. This chapter generalizes the multistate covariate dependent Markov models to higher orders. It may be noted here that the higher order mult-state Markov models are generalization of higher order Markov models for binary outcomes as well. In many situations, multistate models of higher order are useful for examining the underlying characteristics or patterns that prevail in a problem. Some common examples can be given from the problems in the process of a chronic disease with multiple states of progression, reliability aspects over time with multiple modes of transitions among several states, socio-economic problems with multiple transient outcomes, etc.

8.2. HIGHER ORDER MODEL

Let us consider a stationary process $(Y_{i1}, Y_{i2}, ..., Y_{ij})$ for subject i (i= 1,.2,…,n) at follow-up j (j=1,2,…, n_i). Here Y_{ij} is the response at time t_{ij}. As defined in the previous sections, the past history of subject i at follow-up j is denoted by $H_{ij} = \{Y_{ik}, \text{k=1,2,...,j-1}\}$. The transition models for which the conditional distribution of Y_{ij} given H_{ij} depends on r prior observations $Y_{ij-1}, ..., Y_{i,j-r}$ is considered as the model of order r.

The multiple outcomes are defined by y_{ij}=s, s=0, 1, 2,…, k-1 if an event of level s occurs for the i-th subject at the j-th follow-up where y_{ij}=0 indicates that no event occurs. Then the r-th order Markov model can be expressed as

$$P(Y_{ij} \Big| H_{ij}) = P(Y_{ij} \Big| Y_{i,j-r}, ..., Y_{i,j-1}).$$

The probability of a transition from $u_1,...,u_r$ ($u_1,...,u_r$ =0,…,k-1) at times $t_{j-1},...,t_{j-r}$ respectively to s (s=0,…,k-1) at time t_j is

$$\pi_{u_r....u_1 s} = P(Y_{i,j} = s \Big| Y_{i,j-r} = u_r, ..., Y_{i,j-1} = u_1).$$

It is evident that for any combination of $u_r....u_1$, $\sum_{s=0}^{k-1} \pi_{u_r...u_1 s} = 1$, $u_1,...,u_r$ =0,…,k-1 .

Define the following notations:

$X_{ij} = \left[1, \ X_{ij1}, \ ...., X_{ijp}\right]$ = vector of covariates for the i-th person at the (j-r)-th follow-up;

$\beta_{msu} = \left[\beta'_{ms0}, \beta'_{ms1}, \ ...., \beta'_{msp}\right]'$ = vector of parameters for the transition type $u_1...u_r$ to s (m indicates one of the k^r paths $u_r,...,u_1$).

We can express the transition probabilities from state $u_1...u_r$ to state s as follows in terms of conditional probabilities:

$$\pi_{u_1..u_r s}(Y_{i,j} = s \Big| Y_{i,j-r} = u_r, ..., Y_{i,j-1} = u_1, X_{ij}) =$$

$$\pi_{ms}(Y_{i,j} = s \Big| Y_{i,j-r} = u_r, ..., Y_{i,j-1} = u_1, X_{ij}) = \frac{e^{g_{ms}(X_{ij})}}{\sum_{s=0}^{k-1} e^{g_{ms}(X_{ij})}}, \quad (8.1)$$

$$u_1, ..., u_r = 0, ..., k-1$$

where

$$g_{ms}(X_{ij}) = \begin{cases} 0, & \text{if } s = 0 \\ \ln\left[\dfrac{\pi_{u_1...u_r s}(Y_{i,j} = s| \ Y_{i,j-r} = u_r, ..., Y_{i,j-1} = u_1, X_{ij})}{\pi_{u_1...u_r s}(Y_j = 0| \ Y_{i,j-r} = u_r, ..., Y_{i,j-1} = u_1, X_{ij})}\right], & \text{if } s = 1,...,k\text{-}1 \end{cases}$$

and

$$g_{ms}(X_{ij}) = \beta_{ms0} + \beta_{ms1} X_{ij1} + ... + \beta_{msp} X_{ijp}.$$

Then from (8.1) the likelihood function for n individuals with i-th individual having n_i (i=1,2,….n) follow-ups can be expressed as

$$L = \prod_{m=1}^{k^r} \prod_{s=0}^{k-1} \prod_{i=1}^{n} \prod_{j=1}^{n_i} \left[\left\{\pi_{msij}\right\}^{\delta_{msij}}\right] \quad (8.2)$$

where δ_{msij}=1 if a transition type $u_1 \rightarrow ... \rightarrow u_r \rightarrow s$ is observed during j-th follow-up for the i-th individual, $\delta_{msij} = 0$, otherwise, $u_1,...,u_r$, s=0,…,k-1. From (8.2) the log likelihood function is

$$\ln L = \sum_{m=1}^{k^r} \ln L_m ,$$

where L_m corresponds to the $u_1...u_r$ -th component of the likelihood function.

Hence,

$$\ln L_m = \sum_{s=0}^{k-1}\sum_{i=1}^{n}\sum_{j=1}^{n_i}[\delta_{msij} g_{ms}(X_{ij}) - \ln\left(\sum_{s=0}^{k-1} e^{g_{ms}(X_{ij})}\right)].$$

Differentiating $\ln L$ with respect to the parameters and setting the derivatives zero we obtain the following equations

$$\frac{\partial \ln L_m}{\partial \beta_{msq}} = \sum_{i=1}^{n}\sum_{j=1}^{n_i} X_{qij}(\delta_{msij} - \pi_{msij}) = 0 , \tag{8.3}$$

s=0,1,…,k-1; q=0,1,2,….,p; $u_1,...,u_r$=0,…,k-1.

Solving (8.3) we get the likelihood estimates for (k-1)(p+1) parameters.

The second derivatives of the log-likelihood are

$$\frac{\partial^2 \ln L_m}{\partial \beta_{msq} \partial \beta_{msq'}} = -\sum_{i=1}^{n}\sum_{j=1}^{n_i} X_{q'ij} X_{qij} \pi_{msij}(1 - \pi_{msij}),$$

q, q'=0,1,2,….,p; s=0,…,k-1; $u_1,...,u_r$=0,…,k-1.

and

$$\frac{\partial^2 \ln L_m}{\partial \beta_{msq} \partial \beta_{ms'q'}} = -\sum_{i=1}^{n}\sum_{j=1}^{n_i} X_{q'ij} X_{qij} \pi_{msij} \pi_{ms'ij} ,$$

q, q'=0,1,2,….,p; s, s'=0,…,k-1; $u_1,...,u_r$=0,…,k-1. (8.4)

The observed information matrix can be obtained from (8.4).

8.3 Tests for the Model and Parameters

The vectors of $k^r(k-1)$ sets of parameters for the r-th order Markov model can be represented by the following vector:

$$\beta = \left[\beta_1', \beta_2', ..., \beta_{k^r(k-1)}'\right]'$$

where $\beta_v' = \left[\beta_{v0}, \beta_{v1}, ..., \beta_{vp}\right]$, v=1,2,….., $k^r(k-1)$.

To test the null hypothesis H_0 about joint vanishing of all regression parameters (excluding intercepts), we can employ the usual likelihood ratio test

$$-2[\ln L(\beta_0) - \ln L(\beta)] \approx \chi^2_{k^r(k-1)p} \qquad (8.5)$$

For a first order Markov model with dichotomous transition outcomes with p independent variables, the likelihood ratio chi square (8.5) reduces to

$$-2[\ln L(\beta_0) - \ln L(\beta)] \sim \chi^2_{2^1(2-1)p} = \chi^2_{2p}. \qquad (8.6)$$

Similarly, for a second order model with binary outcomes with p independent variables, the null hypothesis of null regression parameter vector can be tested by using the following test statistic:

$$-2[\ln L(\beta_0) - \ln L(\beta)] \sim \chi^2_{2^2(2-1)p} = \chi^2_{4p}. \qquad (8.7)$$

If the number of states for the outcome variable is three and the number of independent variables is p in a first order Markov model, then the chi square can be obtained as follows for testing the null hypothesis of regression parameter vector

$$-2[\ln L(\beta_0) - \ln L(\beta)] \sim \chi^2_{3^1(3-1)p} = \chi^2_{6p}. \qquad (8.8)$$

For a second order model with k=3 and p independent variables, the test statistic is chi square with 18p degrees of freedom as

$$-2[\ln L(\beta_0) - \ln L(\beta)] \sim \chi^2_{3^2(3-1)p} = \chi^2_{18p}. \qquad (8.9)$$

Note that (8.6), (8.7), (8.8) and (8.9) are all special cases of (8.5).

To test the significance of the q-th (q=0,1,...,p) parameter of the v-th (v=1,2,..., $k^r(k-1)$) set of parameters, the null hypothesis is $H_0 : \beta_{vq} = 0$ and the corresponding Wald test is

$$W = \hat{\beta}_{vq} / se(\hat{\beta}_{vq}).$$

8.4 Applications

The HRS data on mobility of elderly population for the period 1992-2004 is used in this chapter as well. We have defined the outcome variable based on the difficulty in mobility of elderly population. We have considered 0= no difficulty, 1= difficulty in one of the five tasks, 2= difficulty in performing two or more tasks. We have considered a second order Markov model for three states. Table 8.1 displays the pooled number of transitions during two consecutive follow-ups in the period 1992-2004 for a three state Markov model. If we examine transition probabilities based on previous two states then 0-0 type results in the largest probability of transition (0.847) for 0-0-0, similarly, 0-1 type results in 0.477 for 0-1-0, 0-2 type shows the largest probability for 0-2-2 (0.471). Similarly, the largest probabilities are observed in each row for 1-0-0 (0.553), 1-1-1 (0.477), 1-2-2 (0.630), 2-0-0 (0.431), 2-1-2

(0.497), and 2-2-2 (0.829). The general pattern is to stay in the states of occupancy in the just prior states with some exceptions.

Table 8.1: Transition Counts and Transition Probability Matrix for the Second Order Three State Mobility Index

Mobility Index	Transition Count			Transition Probability			Total Count
	0	1	2	0	1	2	
00	15687	2010	823	0.847	0.109	0.044	18520
01	1470	1100	511	0.477	0.357	0.166	3081
02	355	350	628	0.266	0.263	0.471	1333
10	1225	654	336	0.553	0.295	0.152	2215
11	630	1128	608	0.266	0.477	0.257	2366
12	199	368	966	0.130	0.240	0.630	1533
20	310	191	218	0.431	0.266	0.303	719
21	209	352	554	0.187	0.316	0.497	1115
22	225	517	3605	0.052	0.119	0.829	4347

The null hypothesis is $H_0 : \pi_{ms} = \pi_{ms}^0$, where m stands for two previous state occupancies such as m=1 (0-0), 2 (0-1), 3 (0-2), 4 (1-0), 5 (1-1), 6 (1-2), 7 (2-0), 8 (2-1), 9 (2-2) and s represents current occupancy of states (s=0,1,2) and the test statistic for a three state Markov model is

$$\chi^2 = \sum_{m=1}^{3^2} \sum_{s=0}^{3-1} \frac{n_{m.}(\hat{\pi}_{ms} - \pi_{ms}^0)^2}{\pi_{ms}^0}$$

which is χ^2 with 9(3-1)-d degrees of freedom under H_0 where d is the number of zeros in π^0. If we set the null hypothesis value of π_{ms}^0=1/3, then the chi square value is 29115.840 (p-value=0.000). Hence we may reject the null hypothesis that π_{ms}^0=1/3.

Similarly, the likelihood ratio test for a second order three state model is

$$-2\ln\Lambda = 2\sum_{m=1}^{3^2} \sum_{s=0}^{3-1} n_{ms} \ln\frac{n_{ms}}{n_{m.}\pi_{ms}^0}.$$

This is χ^2 with 9(3-1) degrees of freedom under H_0. Table 8.2 shows for π_{ms}^0=1/3 that the chi square value is 28268.398, which is quite large. The null hypothesis may be rejected (p-value=0.000).

Table 8.2. Test for the Transition Probabilities of the Second Order Three State Mobility Index

Test-statistics	Value	D.F.	p-value
Chi-square	29115.8401	18	0.000000
LRT	28268.3987	18	0.000000

The test procedure for second order three state Markov model is illustrated below. The results are presented in Table 8.3.

(i) Test for Stationarity Based on Pooled Transition Data

The null hypothesis is $H_0 : \pi_{ms}^t = \pi_{ms}$ and the likelihood ratio chi square is

$$-2\ln\Lambda = 2\sum_{t=1}^{T-2}\sum_{m=1}^{3^2}\sum_{s=0}^{3-1} n_{ms}^t \ln\frac{n_{ms}^t}{n_{m.}^{t-1}\pi_{ms}}.$$

where π_{ms} represents the transition probabilities for the pooled time, T=7.

This is χ^2 with (7-2)9(3-1)=90 degrees of freedom under H_0. Table 8.3 shows that the value of chi square for the data under study is 127.863 and the corresponding p value is less than 0.01. Hence, the null hypothesis of stationarity can be rejected for the second order three state Markov model for the given data.

(ii) Test for Stationarity Based on Consecutive Follow-up Data

The null hypothesis is $H_0 : \pi_{ms}^t = \pi_{ms}^{t+1}$ where

π_{ms}^t = transition probability from m to s for order 2 at time t and

π_{ms}^{t+1} = transition probability from m to s for for order 2 at time t+1

then the likelihood ratio test statistic is

$$-2\ln\Lambda = 2\sum_{t=1}^{T-3}\sum_{m=1}^{3^2}\sum_{s=0}^{3-1} n_{ms}^t \ln\frac{n_{ms}^t}{n_{m.}^{t-1}\pi_{ms}^{t+1}}.$$

This is χ^2 with (7-3)9(3-1)=72 degrees of freedom under H_0. The chi square based on the consecutive follow-up data is 220.902. The corresponding p value for chi square with 72 degrees of freedom is less than 0.001. The null hypothesis of stationarity for the second order three state Markov model may be rejected under this test procedure as well.

Table 8.3. Stationarity Test for the Second Order Three State Mobility Index

Time	D.F.	Comparison between Consecutive Time Points		Comparison with Pooled Transition Matrix	
		Chi-square	p-value	Chi-square	p-value
3	18	84.814384	0.000000	29.901295	0.038417
4	18	45.207600	0.000387	31.354179	0.026186
5	18	65.707112	0.000000	23.615138	0.168045
6	18	25.173109	0.120222	22.833023	0.197088
7	18			20.159860	0.323892
Overall χ^2	72,90	220.90221	0.000000	127.86350	0.005385

Table 8.4 shows the estimates of parameters for the 18 models for the second order transition probabilities for mobility index of the elderly population. The dependent variable is defined to have three states: 0= no difficulty, 1= difficulty in one of the five tasks, 2= difficulty in performing two or more tasks. Age and gender appear to have significant association with 0-0-1 type of transition, age with positive and gender having negative relationship. However, no such associations have been observed between selected variables and 0-0-2 type of transition. The reverse transition of the type 0-1-0 does not show any significant association with the selected variables but progression of the type 0-1-2 is positively associated with age. Age is positively associated with both 0-2-0 and 0-2-1 and gender shows positive relationship with 0-2-0 type of reverse transition. Gender has negative association with reverse transition of the type 1-0-1 but does not show any such association with 1-0-2. Gender increases the likelihood of transition to 1-1-0 but does not show any association with 1-1-2. Age decreases the risk of transition of the type 2-0-1 and 2-0-2, and in addition, gender decreases the transition of the type 2-0-1. Gender is positively and White race as compared to other race is negatively associated with 2-1-0.

Table 8.4. Estimates of the Parameters of the Logistic Regression Models for the Three State Second Order Markov Model for Mobility Index

Variables	Coeff.	Std. error	t-value	p-value	95% C. I.	
					LL	UL
$0 \rightarrow 0 \rightarrow 1$						
Constant	-3.018	0.357	-8.459	0.000	-3.717	-2.318
Age	0.025	0.006	4.368	0.000	0.014	0.036
Gender	-0.430	0.048	-9.050	0.000	-0.524	-0.337
White	-0.175	0.130	-1.340	0.180	-0.430	0.081
Black	0.137	0.141	0.970	0.332	-0.140	0.413
$0 \rightarrow 0 \rightarrow 2$						
Constant	-1.040	0.702	-1.480	0.139	-2.416	0.337
Age	-0.006	0.011	-0.500	0.617	-0.028	0.016
Gender	0.016	0.099	0.161	0.872	-0.178	0.210
White	-0.139	0.259	-0.535	0.593	-0.647	0.369
Black	0.056	0.278	0.201	0.841	-0.490	0.602
$0 \rightarrow 1 \rightarrow 0$						
Constant	-1.576	0.931	-1.692	0.091	-3.400	0.249
Age	0.014	0.015	0.960	0.337	-0.015	0.044
Gender	0.110	0.125	0.875	0.382	-0.136	0.355
White	-0.268	0.321	-0.835	0.403	-0.898	0.361
Black	-0.302	0.343	-0.881	0.378	-0.975	0.370
$0 \rightarrow 1 \rightarrow 2$						
Constant	-4.042	0.545	-7.422	0.000	-5.109	-2.974
Age	0.022	0.009	2.595	0.009	0.005	0.039
Gender	-0.109	0.071	-1.531	0.126	-0.249	0.031
White	-0.008	0.217	-0.038	0.970	-0.432	0.416
Black	0.407	0.229	1.781	0.075	-0.041	0.855

Table 8.4. Continued

Variables	Coeff.	Std. error	t-value	p-value	95% C. I.	
					LL	UL
0 → 2 → 0						
Constant	-1.165	0.526	-2.214	0.027	-2.196	-0.133
Age	0.032	0.008	3.847	0.000	0.016	0.049
Gender	0.189	0.074	2.545	0.011	0.043	0.334
White	-0.157	0.199	-0.789	0.430	-0.547	0.233
Black	-0.384	0.214	-1.792	0.073	-0.804	0.036
0 → 2 → 1						
Constant	-1.467	0.940	-1.561	0.119	-3.309	0.375
Age	0.029	0.015	1.917	0.055	-0.001	0.058
Gender	-0.025	0.127	-0.194	0.846	-0.274	0.225
White	-0.119	0.334	-0.355	0.723	-0.774	0.536
Black	-0.537	0.358	-1.502	0.133	-1.239	0.164
1 → 0 → 1						
Constant	-1.144	0.706	-1.621	0.105	-2.528	0.239
Age	0.016	0.011	1.413	0.158	-0.006	0.038
Gender	-0.215	0.098	-2.199	0.028	-0.406	-0.023
White	-0.135	0.256	-0.528	0.598	-0.638	0.367
Black	-0.084	0.278	-0.301	0.763	-0.628	0.461
1 → 0 → 2						
Constant	-1.124	0.740	-1.519	0.129	-2.575	0.326
Age	0.011	0.012	0.940	0.347	-0.012	0.034
Gender	0.100	0.100	1.000	0.317	-0.096	0.296
White	-0.172	0.280	-0.615	0.538	-0.721	0.377
Black	0.265	0.298	0.889	0.374	-0.319	0.849
1 → 1 → 0						
Constant	-2.235	1.155	-1.935	0.053	-4.498	0.029
Age	0.006	0.018	0.303	0.762	-0.030	0.042
Gender	0.351	0.157	2.239	0.025	0.044	0.659
White	0.029	0.414	0.069	0.945	-0.783	0.840
Black	0.207	0.435	0.475	0.635	-0.646	1.060
1 → 1 → 2						
Constant	-1.904	0.903	-2.108	0.035	-3.674	-0.133
Age	0.004	0.014	0.255	0.798	-0.024	0.032
Gender	-0.035	0.123	-0.287	0.774	-0.277	0.206
White	0.136	0.346	0.391	0.696	-0.543	0.815
Black	0.236	0.370	0.637	0.524	-0.489	0.960
1 → 2 → 0						
Constant	0.612	0.727	0.841	0.400	-0.814	2.038
Age	-0.019	0.011	-1.640	0.101	-0.041	0.004
Gender	0.043	0.099	0.431	0.666	-0.152	0.237
White	-0.119	0.271	-0.439	0.661	-0.651	0.413
Black	-0.186	0.292	-0.635	0.525	-0.759	0.387

Table 8.4. Continued

Variables	Coeff.	Std. error	t-value	p-value	95% C. I.	
					LL	UL
1 → 2 → 1						
Constant	0.137	0.905	0.152	0.880	-1.636	1.910
Age	-0.014	0.014	-0.962	0.336	-0.042	0.014
Gender	-0.215	0.130	-1.654	0.098	-0.469	0.040
White	0.107	0.323	0.331	0.740	-0.527	0.741
Black	-0.235	0.345	-0.681	0.496	-0.912	0.442
2 → 0 → 1						
Constant	3.150	1.332	2.365	0.018	0.539	5.760
Age	-0.057	0.021	-2.660	0.008	-0.098	-0.015
Gender	-0.363	0.174	-2.088	0.037	-0.704	-0.022
White	0.573	0.437	1.310	0.190	-0.284	1.430
Black	0.356	0.466	0.763	0.445	-0.558	1.269
2 → 0 → 2						
Constant	2.426	0.949	2.558	0.011	0.567	4.285
Age	-0.031	0.015	-2.079	0.038	-0.061	-0.002
Gender	-0.033	0.127	-0.260	0.795	-0.283	0.216
White	0.236	0.303	0.777	0.437	-0.358	0.829
Black	0.267	0.325	0.821	0.412	-0.370	0.903
2 → 1 → 0						
Constant	-0.959	1.066	-0.900	0.368	-3.048	1.130
Age	-0.019	0.017	-1.103	0.270	-0.053	0.015
Gender	0.614	0.138	4.436	0.000	0.343	0.885
White	-0.782	0.292	-2.680	0.007	-1.354	-0.210
Black	-0.458	0.309	-1.483	0.138	-1.063	0.147
2 → 1 → 2						
Constant	-1.607	1.270	-1.265	0.206	-4.096	0.883
Age	0.023	0.020	1.128	0.259	-0.017	0.063
Gender	0.057	0.164	0.350	0.727	-0.265	0.380
White	-0.616	0.362	-1.702	0.089	-1.326	0.093
Black	-0.416	0.388	-1.070	0.285	-1.177	0.346
2 → 2 → 0						
Constant	-2.694	1.271	-2.120	0.034	-5.185	-0.204
Age	0.008	0.019	0.402	0.688	-0.030	0.046
Gender	0.104	0.161	0.647	0.518	-0.212	0.421
White	0.946	0.530	1.784	0.074	-0.093	1.986
Black	0.790	0.553	1.428	0.153	-0.294	1.874
2 → 2 → 1						
Constant	-1.784	0.732	-2.437	0.015	-3.218	-0.349
Age	0.000	0.012	0.008	0.993	-0.023	0.023
Gender	0.037	0.100	0.373	0.709	-0.158	0.233
White	0.018	0.226	0.081	0.936	-0.425	0.462
Black	-0.051	0.240	-0.212	0.832	-0.521	0.419
Global Chi-square			15542.96	(0.000)		
LRT			19838.28	(0.000)		

8.5 SUMMARY

In Chapter 7, the Markov models of first order with more than two states are described. This chapter displays the generalization of the methods in the previous chapter for higher order. Both the estimation and test procedures are discussed. In addition, examples are provided for the HRS data on mobility index of the elderly population. The tests for specified hypothesis on the transition probabilities and stationarity are extended for higher order multistate Markov models. This generalization will be helpful for the readers dealing with several states and higher order where first order results can be obtained as special case of Chapter 8 results.

Chapter 9

AN ALTERNATIVE FORMULATION BASED ON CHAPMAN-KOLMOGOROV EQUATION

9.1 INTRODUCTION

In many instances, we have to deal with unequal intervals as well as remaining in the same state for longer time before making any transition. In the previous chapters, we have discussed the background theory with applications where we need to observe the status of transition at each follow-up. In some cases, we may skip the individuals who remain in the same state before making any transition at a later stage. This chapter shows the theoretical perspectives along with its applications to real life situations when we may consider only steps prior to transition. The Chapman-Kolmogorov equations can be employed in order to demonstrate the transitions in n+m steps. Based on the Chapman-Kolmogorov equations, the proposed model introduces an improvement over the previous methods in handling runs of events which is common in longitudinal data. It is noteworthy that by using the Chapman-Kolmogorov equations, without loss of generality, we can express the conditional probabilities in terms of the transition probabilities generated from Markovian assumptions.

9.2 FORMULATION

Let a stationary process $(y_{i1}, y_{i2}, ..., y_{ij})$ represent the past and present responses for subject i (i= 1,.2,…,n) at follow-up j (j=1,2,…, J_i). Here y_{ij} is the response at time t_{ij}. We can think of y_{ij} as an explicit function of past history of subject i at follow-up j denoted by $H_{ij} = \{y_{ik}, \text{ k=1,2,...,j-1}\}$. The transition models for which the conditional distribution of y_{ij} given H_{ij} depends on r prior observations $y_{ij-1}, ..., y_{ij-r}$, is considered as the model of order r.

The multiple outcomes are defined by y_{ij}=s, s=0,1,2,…,m-1 if an event of level s occurs for the ith subject at the jth follow-up where y_{ij}=0 indicates that no event occurs. Then the first order Markov model can be expressed as

$$P(y_{ij} \middle| H_{ij}) = P(y_{ij} \middle| y_{ij-1}). \tag{9.1}$$

The probability of a transition from u (u=0,1,…,m-1) at time t_{j-1} to v (v=0,1,…,m-1) at time t_j is $\pi_{v|u} = P(Y_j = v|\ Y_{j-1} = u)$. For any u,

$$\sum_{v=0}^{m-1} \pi_{v|u} = 1,\ \text{u=0,1,…, m-1.}$$

We can define the following probabilities for m=3, using the Chapman-Kolmogorov equations and also using equation (9.1). The probability of a transition from u (u=0,…,m- 1) at time t_{j_1-1} to v (v=0,…,m-1) at time t_{j_1} is, where t_{j_1-1} is the time of follow-up just prior to t_{j_1}:

$$\pi_{uv} = P(Y_{j_1} = v, Y_{j_1-1} = u) = P(Y_{j_1} = v|\ Y_{j_1-1} = u).P(Y_{j_1-1} = u). \tag{9.2}$$

The probability of a transition from u (u=0,1,…,m-1) at time t_{j_1-1} (just prior to the follow-up at time t_{j_1}) to v (v=0,1,…,m-1) at time t_{j_2-1} (just prior to the follow-up at time t_{j_2}) and w at time t_{j_2} ($j_2 > j_1$) is

$$\begin{aligned}\pi_{uvw} &= P(Y_{j_2} = w, Y_{j_1-1} = u, Y_{j_2-1} = v)\\ &= P(Y_{j_2} = w \mid Y_{j_1-1} = u, Y_{j_2-1} = v)P(Y_{j_1-1} = u, Y_{j_2-1} = v)\\ &= P(Y_{j_2} = w \mid Y_{j_2-1} = v)P(Y_{j_2-1} = v \mid Y_{j_1-1} = u)P(Y_{j_1-1} = u).\end{aligned} \tag{9.3}$$

Similarly, by (9.1) the probability of a transition from u (u=0,…,m-1) at time t_{j_1-1} (just prior to the follow-up at time t_{j_1}) to v (v=0,1,…,m-1) at time t_{j_2-1} (just prior to the follow-up at time t_{j_2}) to w at time t_{j_3-1}(just prior to the follow-up at time t_{j_3}) and s at time t_{j_3}($j_3 > j_2 > j_1$)is:

$$\begin{aligned}\pi_{uvws} &= P(Y_{j_3} = s, Y_{j_1-1} = u, Y_{j_2-1} = v, Y_{j_3-1} = w)\\ &= P(Y_{j_3} = s \mid Y_{j_1-1} = u, Y_{j_2-1} = v, Y_{j_3-1} = w)P(Y_{j_1-1} = u, Y_{j_2-1} = v, Y_{j_3-1} = w)\\ &= P(Y_{j_3} = s \mid Y_{j_3-1} = w)P(Y_{j_3-1} = w \mid Y_{j_2-1} = v)\\ &\quad (Y_{j_2-1} = v \mid Y_{j_1-1} = u)P(Y_{j_1-1} = u).\end{aligned} \tag{9.4}$$

It is observed that $\pi_{uv}, \pi_{uvw}, \pi_{uvws}$ given in (9.2), (9.3) and (9.4) are initially first, second and third order joint probabilities respectively but later these are expressed in terms of first order Markov probabilities described below:

$$\pi_{v|u} = P(Y_{j_1} = v \mid Y_{j_1 - 1} = u), \tag{9.5}$$

$$\begin{aligned}\pi_{w|uv} &= P(Y_{j_2} = w \mid Y_{j_1 - 1} = u, Y_{j_2 - 1} = v) \\ &= P(Y_{j_2} = w \mid Y_{j_2 - 1} = v) P(Y_{j_2 - 1} = v \mid Y_{j_1 - 1} = u) \\ &= \pi_{w|v} . \pi_{v|u} ,\end{aligned} \tag{9.6}$$

$$\begin{aligned}\pi_{s|uvw} &= P(Y_{j_3} = s \mid Y_{j_1 - 1} = u, Y_{j_2 - 1} = v, Y_{j_3 - 1} = w) \\ &= P(Y_{j_3} = s \mid Y_{j_3 - 1} = w) P(Y_{j_3 - 1} = w \mid Y_{j_2 - 1} = v) \\ &\quad P(Y_{j_2 - 1} = v \mid Y_{j_1 - 1} = u) \\ &= \pi_{s|w} . \pi_{w|v} . \pi_{v|u} .\end{aligned} \tag{9.7}$$

In the conditional probabilities (9.5), (9.6) and (9.7) we have considered that once a transition is made from u to v, then the time of event u will remain fixed for all other subsequent transitions. Here a transition from u to v can happen in the second follow-up or the process can remain in the same state u in consecutive follow-ups before making a transition to v. Similarly, in case of a transition from v to w, the last observed time in state v, before making a transition to w, will remain fixed for any subsequent transition. In other words, we can allow the process to stay in the same state v in consecutive follow-ups prior to making any transition. Finally, if a transition is made from w to s then the process is observed at the last time point in the state of w, before making a transition to s. Here the time of last time of observing w can be different from the occurrence of w for the first time as found in expressions for $\pi_{w|uv}$ (for the first observed time to transition to w and last observed times for u and v) and $\pi_{s|uvw}$ (for the first observed time to transition to s and last observed times for u, v and w).

9.3 The Model

Let us define the following notations:

$X_i = [1, X_{i1}, \ldots\ldots, X_{ip}]$ = vector of covariates for the i-th person;

$\beta'_{uv} = [\beta_{uv0}, \beta_{uv1}, \ldots\ldots, \beta_{uvp}]$ = vector of parameters for the transition from u to v,

$\beta'_{vw} = [\beta_{vw0}, \beta_{vw1}, \ldots\ldots, \beta_{vwp}]$ = vector of parameters for the transition from v to w,

and

$\beta'_{ws} = [\beta_{ws0}, \beta_{ws1}, \ldots\ldots, \beta_{wsp}]$ = vector of parameters for the transition from w to s.

For illustration we consider three states, 0,1 and 2, where 0 and 1 are transient and 2 is an absorbing state. The transition probabilities can then be expressed from state u to state v, state v to state w, and state w to state s in terms of conditional probabilities as function of covariates as follows:

$$\pi_{v|u}(X) = P(Y_{j_1} = v|\ Y_{j_1 - 1} = u, X) = \frac{e^{g_{uv}(X)}}{\sum_{k=0}^{2} e^{g_{uk}(X)}}, \text{ u=0, v=0,1,2;}$$

where

$$g_{uv}(X) = \begin{cases} 0, & \text{if v} = 0 \\ \ln\left[\dfrac{P(Y_j = v|\ Y_{j_1 - 1} = u, X)}{P(Y_j = 0|\ Y_{j_1 - 1} = u, X)}\right], & \text{if v} = 1,2. \end{cases}$$

Hence

$$g_{uv}(X) = \beta_{uv0} + \beta_{uv1}X_1 + ... + \beta_{uvp}X_p.$$

Similarly,

$$\pi_{w|v}(X) = P(Y_{j_2} = w|\ Y_{j_2 - 1} = v, X) = \frac{e^{g_{vw}(X)}}{\sum_{k=0}^{2} e^{g_{vk}(X)}}, \text{ v=0,1; w=0,1,2;}$$

where

$$g_{vw}(X) = \begin{cases} 0, & \text{if w} = 1 \\ \ln\left[\dfrac{P(Y_{j_2} = w|\ Y_{j_2 - 1} = v, X)}{P(Y_{j_2} = 1|\ Y_{j_2 - 1} = v, X)}\right], & \text{if w} = 0,2. \end{cases}$$

Hence

$$g_{vw}(X) = \beta_{vw0} + \beta_{vw1}X_1 + ... + \beta_{vwp}X_p;$$

and

$$\pi_{s|w}(X) = P(Y_{j_3} = s|\ Y_{j_3 - 1} = w, X) = \frac{e^{g_{ws}(X)}}{\sum_{k=0}^{2} e^{g_{wk}(X)}}, \text{ w=0,1; s=0,1,2;}$$

where

$$g_{ws}(X) = \begin{cases} 0, & \text{if s} = 0 \\ \ln\left[\dfrac{P(Y_{j_3} = s|\ Y_{j_3 - 1} = w, X)}{P(Y_{j_3} = 0|\ Y_{j_3 - 1} = w, X)}\right], & \text{if s} = 1,2. \end{cases}$$

Hence

$$g_{ws}(X) = \beta_{ws0} + \beta_{ws1}X_1 + ... + \beta_{wsp}X_p.$$

9.4 ESTIMATION

Using (9.5)-(9.7) the likelihood function for n individuals with i-th individual having J_i (i=1,2,….n) follow-ups can be expressed as

$$L = \prod_{i=1}^{n} [\prod_{j_1=1}^{j_2} \prod_{u=0} \prod_{v=0}^{2} (\pi_{v|u}(X_i))^{\delta_{ij_1uv}}][\prod_{j_1=1}^{j_2} \prod_{j_2=j_1}^{j_3} \prod_{u=0} \prod_{v=1} \prod_{w=0}^{2} (\pi_{w|uv}(X_i))^{\delta_{ij_1j_2vw}}]$$
$$[\prod_{j_1=1}^{j_2} \prod_{j_2=j_1}^{j_3} \prod_{j_3=j_2}^{J_i} \prod_{u=0} \prod_{v=1} \prod_{w=0} \prod_{s=0}^{2} (\pi_{s|uvw}(X_i))^{\delta_{ij_1j_2j_3ws}}]. \tag{9.8}$$

The likelihood function given by (9.8) can also be expressed as

$$L = \prod_{i=1}^{n} [\prod_{j_1=1}^{j_2} \{ \prod_{u=0} \prod_{v=0}^{2} (\pi_{v|u}(X_i))^{\delta_{ij_1uv}} \} \prod_{j_2=j_1}^{j_3} \{ \prod_{v=0}^{1} \prod_{w=0}^{2} (\pi_{w|v}(X_i))^{\delta_{ij_1j_2vw}} \}$$
$$\prod_{j_3=j_2}^{J_i} \{ \prod_{w=0}^{1} \prod_{s=0}^{2} (\pi_{s|w}(X_i))^{\delta_{ij_1j_2j_3ws}} \}] \tag{9.9}$$

where δ_{ij_1uv}=1 if a transition type $u \to v$ (u=0, v=1,2) is observed at j_1 th follow-up for the ith individual, δ_{ij_1uv}=0, otherwise; $\delta_{ij_1j_2vw}$=1, if a transition type $u \to v$ (u=0, v=1,2) is observed at j_1 th follow-up and a transition type $v \to w$ (v=1, w=0,2) is observed at j_2 th follow-up, $\delta_{ij_1j_2vw}$=0, if a transition type $u \to v$ (u=0, v=1,2) is observed at j_1 th follow-up and a transition type $v \to w$ (v=1, w=0,2) does not occur at j_2 th follow-up; $\delta_{ij_1j_2j_3ws}$=1 if a transition type $u \to v$ (u=0, v=1,2) is observed at j_1 th follow-up, a transition type $v \to w$ (v=1, w=0,2) is observed at j_2 th follow-up, and a transition type $w \to s$ (w=0, s=1,2) is observed at j_3 th follow-up, $\delta_{ij_1j_2vw}$=0, if a transition type $u \to v$ (u=0, v=1,2) is observed at j_1 th follow-up, a transition type $v \to w$ (v=1, w=0,2) is observed at j_2 th follow-up, and a transition type $w \to s$ (w=0, s=1,2) does not occur at j_3 th follow-up.

Taking log of (9.9) the log likelihood function is given by

$$\ln L = \sum_{i=1}^{n} [\sum_{j_1=1}^{J_2} \{[\sum_{u=0} \sum_{v=0}^{2} \delta_{ij_1uv} \ln \pi_{v|u}(X_i)\} + \sum_{j_2=j_1}^{j_3} \{ \sum_{v=0}^{1} \sum_{w=0}^{2} \delta_{ij_1j_2vw} \ln \pi_{w|v}(X_i)\}$$
$$+ \sum_{j_3=j_2}^{J_i} \{ \sum_{w=0}^{1} \sum_{s=0}^{2} \delta_{ij_1j_2j_3ws} \ln \pi_{s|w}(X_i)\}]\}]. \tag{9.10}$$

Differentiating (9.10) with respect to the parameters and setting the derivatives zero we obtain the likelihood equations as

$$\frac{\partial \ln L}{\partial \beta_{uvq}} = \sum_{i=1}^{n} \sum_{j_1=1}^{j_2} \delta_{ij_1uv} X_{qi} (1 - \pi_{v|u}(X_i)) = 0 \ ,$$

$$\frac{\partial \ln L}{\partial \beta_{vwq}} = \sum_{i=1}^{n} \sum_{j_1=1}^{j_2} \sum_{j_2=j_1}^{j_3} \delta_{ij_1 j_2 vw} X_{qi} (1 - \pi_{w|v}(X_i)) = 0,$$

$$\frac{\partial \ln L}{\partial \beta_{wsq}} = \sum_{i=1}^{n} \sum_{j_1=1}^{j_2} \sum_{j_2=j_1}^{j_3} \sum_{j_3=j_2}^{J_i} \delta_{ij_1 j_2 j_3 ws} X_{qi} (1 - \pi_{s|w}(X_i)) = 0. \quad (9.11)$$

Solutions of (9.11) give the MLE's of parameters. The second derivatives of the log likelihood are given by

$$\frac{\partial^2 \ln L}{\partial \beta_{uvq} \partial \beta_{uvq'}} = -\sum_{i=1}^{n} \sum_{j_1=1}^{j_2} \delta_{ij_1 uv} X_{q'i} X_{qi} \pi_{v|u}(X_i)(1 - \pi_{v|u}(X_i)),$$

$$\frac{\partial^2 \ln L}{\partial \beta_{vwq} \partial \beta_{vwq'}} = -\sum_{i=1}^{n} \sum_{j_1=1}^{j_2} \sum_{j_2=j_1}^{j_3} \delta_{ij_1 j_2 vw} X_{q'i} X_{qi} \pi_{w|v}(X_i)(1 - \pi_{w|v}(X_i)),$$

$$\frac{\partial^2 \ln L}{\partial \beta_{wsq} \partial \beta_{wsq'}} = -\sum_{i=1}^{n} \sum_{j_1=1}^{j_2} \sum_{j_2=j_1}^{j_3} \sum_{j_3=j_2}^{J_i} \delta_{ij_1 j_2 j_3 ws} X_{q'i} X_{qi} \pi_{s|w}(X_i)(1 - \pi_{s|w}(X_i)). \quad (9.12)$$

and

$$\frac{\partial^2 \ln L}{\partial \beta_{uvq} \partial \beta_{uv'q'}} = -\sum_{i=1}^{n} \sum_{j_1=1}^{j_2} \delta_{ij_1 uv} X_{q'i} X_{qi} \pi_{v|u}(X_i) \pi_{v'|u}(X_i),$$

$$\frac{\partial^2 \ln L}{\partial \beta_{vwq} \partial \beta_{vw'q'}} = -\sum_{i=1}^{n} \sum_{j_1=1}^{j_2} \sum_{j_2=j_1}^{j_3} \delta_{ij_1 j_2 vw} X_{q'i} X_{qi} \pi_{w|v}(X_i) \pi_{w'|v}(X_i),$$

$$\frac{\partial^2 \ln L}{\partial \beta_{wsq} \partial \beta_{ws'q'}} = -\sum_{i=1}^{n} \sum_{j_1=1}^{j_2} \sum_{j_2=j_1}^{j_3} \sum_{j_3=j_2}^{J_i} \delta_{ij_1 j_2 j_3 ws} X_{q'i} X_{qi} \pi_{s|w}(X_i) \pi_{s'|w}(X_i).$$

(9.13)

The observed information matrix can be obtained from (9.12) and (9.13).

9.5 Tests for the Model and Parameters

For illustrating the test procedure, let us suppose that all the individuals were in state 0 initially. Let us consider here two possible situations: (A) there are two transient states 0 and 1, and (B) there are two transient states 0 and 1 and one absorbing state 2.

Case (A)

In case (A), we may consider the following possible transition types:
(i) u=0, v=1 (uv), (ii) u=0, v=1, w=0 (vw), (iii) u=0, v=1, w=0, s=1 (ws).

If we consider p variables for each model, then $\beta = [\beta_1, \beta_2', \beta_3']'$ where

$\beta_k' = [\beta_{k0}, \beta_{k1}, ..., \beta_{kp}]$, k=1,2,3. Then the likelihood ratio chi square for testing joint vanishing of all regression parameter (excluding intercepts) is

$$-2[\ln L(\beta_0) - \ln L(\beta)] \sim \chi^2_{3p}.$$

<u>Case (B)</u>

In case (B), we have considered two transient states 0 and 1 and one absorbing state 2. Then the transition types are:

(i) u=0, v=1 (uv), (ii) u=0, v=2 (uv), (iii) u=0, v=1, w=0 (vw), (iv) u=0, v=1, w=2 (vw), (v) u=0, v=1, w=0, s=1 (ws), and (vi) u=0, v=1, w=0, s=2 (ws).

In other words, we have to fit models for three transition types for case (A) and models for six transition types for case (B).

If we consider p variables for each model, then $\beta = [\beta_1', \beta_2, \beta_3', \beta_4', \beta_5', \beta_6']'$ where

$\beta_k' = [\beta_{k0}, \beta_{k1}, ..., \beta_{kp}]$, k=1,2,3,4,5,6. Then the likelihood ratio chi square is

$$-2[\ln L(\beta_0) - \ln L(\beta)] \sim \chi^2_{6p}.$$

To test the significance of the qth parameter of the k-th set of parameters for both cases (A) and (B), the null hypothesis is $H_0 : \beta_{kq} = 0$ and the corresponding Wald test statistic is $W = \hat{\beta}_{kq} / se(\hat{\beta}_{kq})..$

9.6 Application

For illustration, we have used the mental health index that was derived using a score on the Center for Epidemiologic Studies Depression (CESD) scale. The CESD score is the sum of eight indicators (ranges 0 to 8). The negative indicators measure whether the respondent experienced the following sentiments all or most of the time: depression, everything is an effort, sleep is restless, felt alone, felt sad, and could not get going. The positive indicators measure whether the respondent felt happy and enjoyed life, all or most of the time. These two were reversed before adding in the score. The score ranges from 0 to 8. We have categorized this into two categories: 0 for no depression and 1 for mild or severe depression.

Table 9.1 shows the distribution of subjects by transition types. It shows that 46.3 percent moved from no depression to depression (u=0, v=1), 54.5 percent of them experienced reverse transition to no depression (u=0, v=1, w=0), and 51.8 percent of the reverse transitions resulted in repeated transition.

Table 9.1. Distribution of Respondents by Transitions, Reverse Transitions and Repeated Transitions for Mobility Index (Two States)

Transition Types	N	%
Transitions		
No transition from no depression	3731	53.7
No depression to depression (0→1)	3222	46.3
Total	6953	100.0
Reverse Transitions		
No transition after transition to depression	1465	45.5
No depression to depression to no depression (0→1→0)	1757	54.5
Total	3222	100.0
Repeated Transitions		
No transition after reverse transition	847	48.2
No depression to depression to no depression to depression (0→1→0→1)	910	51.8
Total	1757	100.0

The fit of the logistic regression models with covariates age, gender, White and Black races are displayed in Table 9.2. For transition type 0-1, age and gender show negative associations, and both White and Black races, compared to other races, indicate positive associations with transition to depression from no depression. For reverse transition, age shows negative association, and for repeated transition, both age and gender show negative associations.

Table 9.3 displays the distribution of individuals by transition types for case (B). In this case, we have considered three states: 0= no depression, 1=mild or severe depression, and 2= death. It shows that the transitions to depression and death are respectively 48.1 percent and 3.4 percent respectively. After transition to depression, 55.8 percent have experienced reverse transition to no depression and 6.8 percent died. Thereafter, 53.3 percent repeated the transition to depression and 2.3 percent died.

Table 9.2. Estimates of the Parameters of the Logistic Regression Models for Transitions, Reverse Transitions and Repeated Transitions (Two States)

Variables	Coeff.	SE	Wald	p-value	Odds Ratio	95% CI	
						Lower	Upper
Transition 0→1							
Intercept	7.711	0.324	565.206	0.000	2232.631		
Age	-0.133	0.005	701.282	0.000	0.875	.867	.884
Gender	-0.464	0.052	81.147	0.000	0.628	.568	.695
White	0.419	0.140	8.927	0.003	1.520	1.155	2.001
Black	0.436	0.152	8.260	0.004	1.546	1.149	2.081
Reverse Transitions (0→1→0)							
Intercept	12.039	0.603	398.716	0.000	169219.74		
Age	-0.188	0.009	433.334	0.000	0.828	0.814	0.843
Gender	-0.073	0.078	0.878	0.349	0.930	0.798	1.083
White	0.029	0.222	0.017	0.898	1.029	0.666	1.590
Black	-0.253	0.237	1.147	0.284	0.776	0.488	1.234
Repeated Transitions (0→1→0→1)							
Intercept	13.496	0.934	208.955	.000	726438.18		
Age	-0.204	0.014	214.986	.000	0.815	0.793	0.838
Gender	-0.242	0.105	5.324	.021	0.785	0.640	0.964
White	-0.137	0.288	0.225	.636	0.872	0.496	1.535
Black	-0.280	0.311	0.812	.368	0.756	0.411	1.390
Model Chi-square			1740.53	(d.f=12,	p-value=0.000)		

Table 9.3. Distribution of Respondents for Transitions, Reverse Transitions and Repeated Transitions for Multistate Analysis (Three States)

Transition Types	N	%
Transitions		
No transition from no depression	3370	48.5
No depression to depression (0→1)	3345	48.1
No depression to Death (0→2)	238	3.4
Total	6953	100.0
Reverse Transition and Transition to Death		
No transition after transition to depression	1250	37.4
No depression to depression to no depression (0→1→0)	1868	55.8
No depression to depression to death (0→1→2)	227	6.8
Total	3345	100.0
… Repeated Transition and Transition to Death		
No transition after reverse transition	826	44.2
No depression to depression to no depression to depression (0→1→0→1)	999	53.5
No depression to depression to no Depression to Death (0→1→0→2)	43	2.3
Total	1868	100.0

Table 9.4 shows the fit of the logistic regression models for each of the transition types for case (B). For transition type 0-1, age indicates negative association while gender and White and Black races show positive associations. However, transition type 0-2 is negatively associated with both age and gender. With reverse transition 0-1-0, only age shows negative association whereas transition type 0-1-2 is associated negatively with both age and gender. Similarly, age and gender are associated negatively and positively respectively with repeated transition 0-1-0-1 but only age appears to have negative association with transition type 0-1-0-2.

Table 9.4. Estimates of the Parameters of the Proposed Multistate Models for Transitions, Reverse Transitions and Repeated Transitions (Three States)

Variables	Coeff.	SE	Wald	p-value	Odds Ratio	95% CI	
						Lower	Upper
Transition 0→1							
Intercept	8.030	0.345	541.649	0.000			
Age	-0.151	0.005	854.363	0.000	0.860	0.851	0.868
Gender	0.464	0.053	76.071	0.000	1.590	1.433	1.765
White	0.458	0.144	10.117	0.001	1.581	1.192	2.097
Black	0.514	0.156	10.816	0.001	1.671	1.231	2.270
Transition 0→2							
Intercept	4.290	0.847	25.646	0.000			
Age	-0.098	0.013	59.181	0.000	0.907	0.885	0.930
Gender	-0.755	0.151	25.170	0.000	0.470	0.350	0.631
White	0.033	0.340	0.009	0.923	1.033	0.530	2.013
Black	0.396	0.365	1.179	0.277	1.487	0.727	3.040
Reverse Transitions (0→1→0)							
Intercept	15.971	0.702	517.560	0.000			
Age	-0.248	0.010	579.973	0.000	0.780	0.765	0.796
Gender	-0.033	0.084	0.155	0.694	0.967	0.821	1.140
White	0.216	0.238	0.826	0.363	1.241	0.779	1.979
Black	-0.019	0.254	0.005	0.941	0.981	0.597	1.613
Transitions to Death (0→1→2)							
Intercept	11.107	1.183	88.161	0.000			
Age	-0.190	0.017	121.604	0.000	0.827	0.800	0.855
Gender	-0.693	0.150	21.290	0.000	0.500	0.372	0.671
White	0.315	0.457	0.475	0.491	1.370	0.560	3.351
Black	0.508	0.478	1.129	0.288	1.661	0.651	4.237
Repeated Transition (0→1→0→1)							
Intercept	15.420	0.986	244.765	0.000			
Age	-0.241	0.015	274.006	0.000	0.785	0.763	0.808
Gender	0.270	0.105	6.579	0.010	1.310	1.066	1.611
White	0.002	0.296	0.000	0.993	1.002	0.562	1.790
Black	-0.194	0.318	0.372	0.542	0.824	0.442	1.536

Table 9.4. Continued

Variables	Coeff.	SE	Wald	p-value	Odds Ratio	95% CI	
						Lower	Upper
Transition to Death (0→1→0→2)							
Intercept	8.505	2.836	8.994	0.003			
Age	-0.171	0.041	17.175	0.000	0.843	0.777	0.914
Gender	-0.474	0.319	2.209	0.137	0.623	0.333	1.163
White	0.305	1.043	0.085	0.770	1.357	0.176	10.485
Black	0.855	1.075	0.632	0.427	2.351	0.286	19.337
Model Chi-square			2378.51 (d.f=24, p-value=0.000)				

9.7 Summary

The construction of likelihood function for repeated observations poses formidable difficulty if the occurrence or nonoccurrence of event is observed at unequal intervals of duration. For instance, the transition may occur from 0 to 1 during first to second follow-up but then the next event from 1 to 0 may occur between 5th to 6th follow-up. If we consider these events consecutively then the analysis of transitions fails to reflect the overall transition patterns. This chapter uses the Chapman-Kolmogorov equations to provide conditional transition probabilities after a certain number of steps and thus provides more insights to the transitions whenever it happens. In other words, the analysis becomes more specific in terms of relationship between transition probabilities and the underlying covariates.

Chapter 10

ADDITIONAL INFERENCE PROCEDURES

10.1 INTRODUCTION

This chapter provides a summary of the test procedures provided in covariate dependent Markov models of higher order and some additional test procedures for assessing the order of the underlying Markov models. We have already shown that the likelihood function based likelihood ratio tests can be employed for testing the significance of the overall model. In addition, we have employed the Wald test for testing the significance of parameters associated with the potential risk factors of interest. For testing the adequacy of a Markov model, we can consider Crowley-James method or a simple alternative, a method based on regressive logistic regression method (Bonney, 1987) or it's simplified version, etc.

10.2 TESTS FOR THE MODEL

The vectors of $m^r(m-1)$ sets of parameters for the r-th order m-state Markov model can be represented by the following vector:

$$\beta = \left[\beta_1', \beta_2', \ldots\ldots\ldots, \beta'_{m^r(m-1)}\right]'$$

where $\beta_v' = \left[\beta_{v0}, \beta_{v1}, \ldots\ldots, \beta_{vp}\right]$, v=1,2,….., $m^r(m-1)$..

To test the null hypothesis H_0 about the joint vanishing of regression parameter (excluding intercepts) we can employ the usual likelihood ratio test

$$-2[\ln L(\hat{\beta}_0) - \ln L(\hat{\beta})] \sim \chi^2_{m^r(m-1)p} \tag{10.1}$$

where $\hat{\beta}$ is the MLE of β.

For a first order (r=1) Markov model with dichotomous (m=2) transition outcomes with p independent variables (10.1) reduces to

$$-2[\ln L(\hat{\beta}_0) - \ln L(\hat{\beta})] \sim \chi^2_{2^1(2-1)p} = \chi^2_{2p}.$$

Similarly, for a second order (r=2) model with binary outcomes (m=2) with p independent variables, (10.1) reduces to

$$-2[\ln L(\hat{\beta}_0) - \ln L(\hat{\beta})] \sim \chi^2_{2^2(2-1)p} = \chi^2_{4p}.$$

If the number of states for the outcome variable is three (m=3) and the number of independent variables is p in a first order (r=1) Markov model, then (10.1) reduces to

$$-2[\ln L(\hat{\beta}_0) - \ln L(\hat{\beta})] \sim \chi^2_{3^1(3-1)p} = \chi^2_{6p}.$$

For a second order (r=2) model with m=3 states and p independent variables, the (10.1) is chi square with 18p degrees of freedom as shown below:

$$-2[\ln L(\hat{\beta}_0) - \ln L(\hat{\beta})] \sim \chi^2_{3^2(3-1)p} = \chi^2_{18p}.$$

10.3 Test for Individual Parameters

To test the significance of the qth (q=0,1,…,p) parameter of the v-th (v=1,2,…, $m^r(m-1)$) set of parameters, the null hypothesis is $H_0 : \beta_{vq} = 0$ and the corresponding Wald test is

$$W = \hat{\beta}_{vq} / se(\hat{\beta}_{vq}).$$

10.4 Test for Order of the Model

A simple method is proposed here for testing the order of a Markov model with covariate dependence. We consider a general model for order r and take all the past states as covariates. In this test, we assume no interaction between different transition types. In the model, the transition probability takes into account selected covariates and previous transitions are also incorporated as covariates for an r-order Markov model. The model is as follows:

$$\pi_{k1} = P\left(Y_j = s = 1 \mid Y_{j-r} = s_r, \cdots, Y_1 = s_1, X\right) = \frac{e^{\left(\beta' X + \theta_1 Y_{j-1} + \cdots + \theta_r Y_{j-r}\right)}}{1 + e^{\left(\beta' X + \theta_1 Y_{j-1} + \cdots + \theta_r Y_{j-r}\right)}} \quad (10.2)$$

where k denotes the specific sequence of states $(s_r, s_{r-1}, ..., s_1)$ in previous r time points and $\pi_{k0} = 1 - \pi_{k1}$, or generally, π_{ks}, s=0,1, represent the probabilities of transition to state s via k-th type transition sequence, $\beta' = (\beta_0, \beta_1, ..., \beta_p)$, and $X' = (1, X_1, ..., X_p)$.

Using (10.2) the likelihood function is given by

$$L = \prod_{k=1}^{2^r} \prod_{s=0}^{1} \prod_{i=1}^{n_k} \left[\left\{ \pi_{iks} \right\}^{\delta_{iks}} \right] \quad (10.3)$$

where $\delta_{iks} = 1$ if the outcome at time t_j is $Y_j = 1$ for individual i and $\delta_{iks} = 0$ if the outcome at time t_j is $Y_j = 0$ for individual i for the transition type $s_r, s_{r-1}, ..., s_1$ prior to time t_j and n_k denotes the number of subjects experiencing transition type $Y_{j-r} = s_r, Y_{j-(r-1)} = s_{r-1}, ..., Y_{j-1} = s_1$ prior to time t_j.

The estimates of parameters β and $\theta_1, \cdots, \theta_r$ are obtained from the equations

$$\frac{\partial \ln L}{\partial \beta_\ell} = 0, \; \ell = 1,2,\cdots, p$$

$$\frac{\partial \ln L}{\partial \theta_q} = 0, \; q = 1,2,\cdots, r.$$

derived by differentiating log of (10.3)

The test can be performed in two steps: The first step is to test for the significance of the overall model. In that case the null hypothesis is: $H_0 : \beta_1 = ... = \beta_p = 0, \theta_1 = ... = \theta_r = 0$. If this is significant on the basis of the likelihood ratio test

$$-2[\ln L(\hat{\beta}_0) - \ln L(\hat{\beta}, \theta)] \sim \chi^2_{p+r}$$

where $\beta' = (\beta_0, \beta_1, ..., \beta_p)$ and $\theta' = (\theta_1, \theta_2, ..., \theta_r)$ then we conduct the second test for testing the significance of the order. For the r-th order, we can use the likelihood ratio test for testing the null hypothesis $H_0 : \theta_1 = ... = \theta_r = 0$:

$$-2[\ln L(\hat{\beta}) - \ln L(\hat{\beta}, \hat{\theta})] \sim \chi^2_r.$$

We can also use the Wald's test for testing significance of a particular order for $H_0 : \theta_q = 0$, q=1,2,...,r as $W = \hat{\theta}_q / se(\hat{\theta}_q)$.

10.5 Regressive Logistic Model

Bonney (1987) proposed the regressive logistic model where both binary outcomes in previous times as well as covariates can be included. The joint mass function can be expressed as

$$P(y_{i1}, y_{i2}, ..., y_{in_i}; x_i) = P(y_{i1}; x_i) P(y_{i2} | y_{i1}; x_i) P(y_{i3} | y_{i1}; y_{i2}; x_i)...$$

$$P(y_{in_i} | y_{i1}, ..., y_{in_{i-1}}; x_i) \; i = 1,2,...,n. \qquad (10.4)$$

where x_i is the vector of covariate values for i-th individual. The j-th logit is defined as

$$\theta_j = \ln \frac{P(y_{ij} = 1 | y_{i1}, y_{i2}, ..., y_{ij-1}; x_i)}{P(y_{ij} = 0 | y_{i1}, y_{i2}, ..., y_{ij-1}; x_i)}.$$

Bonney proposed regression model for each conditional probability as shown below:

$$P(y_{ij} | y_{i1},...,y_{ij-1};x_i) = \frac{e^{\theta_j y_{ij}}}{1+e^{\theta_j}} = \pi(x_i, y_{i1,...,ij-1}),$$

$$\theta_j = \beta_0 + \beta_1 y_{i1} + ... + \beta_{j-1} y_{ij-1} + \gamma_0 + \gamma_1 x_{i1} + ... + \gamma_p x_{ip}$$

$$= \beta_0^* + \beta_1 y_{i1} + ... + \beta_{j-1} y_{ij-1} + \gamma_1 x_{i1} + ... + \gamma_p x_{ip} \quad (10.5)$$

where $\beta_0^* = \beta_0 + \gamma_0$ and θ_j is the j-th logit as defined above.

Now using (10.4) and (10.5) we can obtain the likelihood function as

$$L = \prod_{i=1}^{n} \prod_{j=1}^{n_i} P(y_{ij} | y_{i0} y_{i1},..., y_{ij-1}; x_i) = \prod_{i=1}^{n} \prod_{j=1}^{n_i} \frac{e^{\theta_j y_{ij}}}{(1+e^{\theta_j})}. \quad (10.6)$$

Taking log of (10.6) we get the log likelihood. The estimates of the parameters can be obtained from the equations of first derivatives of log likelihood function with respect to the parameters contained in θ_j:

$$\frac{\partial \ln L}{\partial \beta} = 0,$$

$$\frac{\partial \ln L}{\partial \gamma} = 0.$$

The elements of the score vector for β are given by

$$\sum_{i=1}^{n} \sum_{j=1}^{n_i} [y_{ij} - \pi(x_{ij}, y_{i1,...,ij-1})] = 0,$$

$$\sum_{i=1}^{n} \sum_{j=1}^{n_i} y_{il} [y_{ij} - \pi(x_{ij}, y_{i1,.}..., y_{ij-1})] = 0 \quad (l = 1,2,...,p). \quad (10.7)$$

The score vector for γ is

$$\sum_{i=1}^{n} \sum_{j=1}^{n_i} [y_{ij} - \pi(x_{ij}, y_{i1,...,ij-1})] = 0,$$

$$\sum_{i=1}^{n} \sum_{j=1}^{n_i} x_{iq} [y_{ij} - \pi(x_{iq}, y_{i1},..., y_{ij-1})] = 0 \quad (q=1,2,...,r). \quad (10.8)$$

Now let us define the set of covariates as follows:

$$\mathbf{z}_{j-1} = (1, y_1,..., y_{j-1}, x_1,..., x_p)'$$

$$= (z_0, z_1,..., z_{j+p-1})'$$

and the corresponding parameters:

$$\boldsymbol{\lambda}_{j-1} = (\beta_0, \beta_1, ..., \beta_{j-1}, \gamma_1, ..., \gamma_p)'$$
$$= (\lambda_0, \lambda_1, ..., \lambda_{j+p-1})$$

In other words, for the outcome variable y_j, there are j+p parameters, where j=1,2,…, n_i.

Hence, we can define $\theta_j = \boldsymbol{\lambda}'_{j-1}\mathbf{z}_{j-1}$.

Then using (10.7) and (10.8) we can rewrite the score vectors as:

$$\sum_{i=1}^{n} \sum_{j=1}^{n_i} z_{i,j-1,q}[y_{ij} - \pi(z_{i,j-1,q})] = 0, \quad \text{q=0,1,…,j+p-1.}$$

The variance and covariance of the estimated coefficients can be obtained from:

$$I^*_{qq'} = -\sum_{i=1}^{n} z_{i,j-1,q} z_{i,j-1,q'} \pi_i (1 - \pi_i)$$

for $q, q' = 0,1,2,..., j + p - 1$ where $\pi_{i,j-1} = \pi(z_{i,j-1})$. Let $I_{qq'} = (-1).I^*_{qq'}$. Then the information matrix is defined by $I(\lambda)$, where (q, q')-th element of $I(\lambda)$ is $I_{qq'}$.

The variances and covariances of the estimated coefficients are obtained from the inverse of the information matrix, ie

$$\Sigma(\lambda) = I^{-1}(\lambda).$$

The information matrix can be expressed as

$$\hat{I}(\hat{\lambda}) = z'Vz$$

where z is an nx(p+1) matrix containing the data for each subject

$$z = \begin{bmatrix} 1 & z_{11} & ....z_{1,j+p-1} \\ 1 & z_{21} & ....z_{2,j+p-1} \\ .... \\ .... \\ 1 & z_{n1} & ....z_{n,j+p-1} \end{bmatrix} \text{ and } V = \begin{bmatrix} \hat{\pi}_1(1-\hat{\pi}_1) & 0............... & 0 \\ 0 & \hat{\pi}_2(1-\hat{\pi}_2)....... & 0 \\ .... \\ .... \\ 0 & 0.............\hat{\pi}_n(1-\pi_n) \end{bmatrix}.$$

10.5.1 Two Response Variables without Covariates

If we have two binary outcome variables, y_1 and y_2 without explanatory variables, then

$$P(y_{i1}, y_{i2}) = P(y_{i1})P(y_{i2}|y_{i1})$$

where

$$P(y_{i1}) = e^{\lambda_0}/(1+e^{\lambda_0}),\ 1 - P(y_{i1}) = 1/(1+e^{\lambda_0}).$$

Alternatively,

$$P(y_{i1}=1)=e^{\lambda_0}/(1+e^{\lambda_0}),\ P(y_{i1}=0)=1/(1+e^{\lambda_0}).$$

Similarly,

$$P(y_{i2}=1|y_{i1}=1)=e^{\lambda_0+\lambda_1}/(1+e^{\lambda_0+\lambda_1}),\ P(y_{i2}=0|y_{i1}=1)=1/(1+e^{\lambda_0+\lambda_1}),$$

$$P(y_{i2}=1|y_{i1}=0)=e^{\lambda_0}/(1+e^{\lambda_0}),\ P(y_{i2}=1|y_{i1}=0)=1/(1+e^{\lambda_0}).$$

The $\lambda's$ in marginal models and conditional models are most likely different, but we have used the same expressions in order to avoid more suffix. Similarly, we can extend it further for any number of dependent binary outcomes.

10.5.2 Two Response Variables with Covariates

Now if we add one covariate, x, in the model, then

$$P(y_{i1},y_{i2}|x_i)=P(y_{i1}|x_i)P(y_{i2}|y_{i1},x_i)$$

where

$$P(y_{i1}=1|x_i)=e^{\lambda_0+\lambda_2 x_i}/(1+e^{\lambda_0+\lambda_2+x_i}),\ P(y_{i1}=0)=1/(1+e^{\lambda_0+\lambda_2 x_i}).$$

Similarly,

$$P(y_{i2}=1|y_{i1}=1,x_i)=\frac{e^{\lambda_0+\lambda_1+\lambda_2 x_i}}{1+e^{\lambda_0+\lambda_1+\lambda_2 x_i}},\ P(y_{i2}=0|y_{i1}=1,x_i)=\frac{1}{1+e^{\lambda_0+\lambda_1+\lambda_2 x_i}}.$$

$$P(y_{i2}=1|y_{i1}=0,x_i)=\frac{e^{\lambda_0+\lambda_2 x_i}}{1+e^{\lambda_0+\lambda_2 x_i}},\ P(y_{i2}=1|y_{i1}=0,x_i)=\frac{1}{1+e^{\lambda_0+\lambda_2 x_i}}.$$

10.5.3 Equal Predictive Effect

For equal predictive effect, we can test the following null hypothesis:

$$\lambda_1=\lambda_2=...=\lambda_k=\lambda \text{ and } S_k=z_1+...+z_k.$$

Then the model is

$$\theta_j=\lambda_0+\lambda S_{j-1}+\lambda_j z_j+...+\lambda_{j+p-1}z_{j+p-1}.$$

10.5.4 Serial Dependence

For a first order dependence, we can express

$$P(y_{i1},y_{i2},...,y_{in_i})=P(y_{i1}|x_{i1})\prod_{j=2}^{n_i}P(y_{ij}|y_{ij-1},x_{ij})$$

where $\theta_j=\boldsymbol{\lambda}'_{j-1}\mathbf{z}_{j-1}=\lambda_0+\lambda_1 z_{i1}+\lambda_2 z_{i2}.$

Similarly, the second order dependence can be shown as follows:

$$P(y_{i1}, y_{i2}, ..., y_{in_i}) = P(y_{i1} | x_{i1}) \prod_{j=3}^{n_i} P(y_{ij} | y_{ij-1}, y_{ij-2}, x_{ij})$$

and we can show that $\theta_j = \boldsymbol{\lambda}'_{j-1}\mathbf{z}_{j-1} = \lambda_0 + \lambda_1 z_{i1} + \lambda_2 z_{i2} + \lambda_3 z_{i3}$.

We can extend the above for the r-th order as

$$P(y_{i1}, y_{i2}, ..., y_{in_i}) = P(y_{i1} | x_{i1}) \prod_{j=3}^{n_i} P(y_{ij} | y_{ij-1}, y_{ij-2}, ..., y_{ij-r}, x_{ij})$$

and we can show that $\theta_j = \boldsymbol{\lambda}'_{j-1}\mathbf{z}_{j-1} = \lambda_0 + \lambda_1 z_{i1} + \lambda_2 z_{i2} + \lambda_3 z_{i3} + ... + \lambda_r z_{ir}$.

The last variable in these equations correspond to the single covariate, x. Without losing any generality, we can extend the same equations for p covariates. In addition, we can also consider the interaction between two, three,..., and r outcome variables prior to j for a fully parameterized model. In that case, the number of classes or cells would be 2^{r+1}, only for highlighting the dependence in outcome variables.

10.5.5 Difference between Bonney's Approach and the Proposed Approach

If we consider equal number of follow-up observations for every individual in the sample, then $n_i = T$, where T is the number of follow-ups. Then Bonney (1987) defined

$$P(y_{i1}, y_{i2}, ..., y_{iT}; x) = P(y_{i1}; x)P(y_{i2} | y_{i1}; x)P(y_{i3} | y_{i1}; y_{i2})...P(y_{iT} | y_{i1}, ..., y_{iT-1}; x).$$

For any j<T, we can denote the conditional probability as $P(y_{ij} | y_{i1}, y_{i2}, ..., y_{i,j-1}; x_i)$.

We can also show that

$$\theta_j = \ln \frac{P(y_{ij} = 1 | y_{i1}, y_{i2}, ..., y_{ij-1}; x_i)}{P(y_{ij} = 0 | y_{i1}, y_{i2}, ..., y_{ij-1}; x_i)} = \boldsymbol{\lambda}'_{j-1}\mathbf{z}_{j-1}$$

where

$$\mathbf{z}_{j-1} = (1, y_1, ..., y_{j-1}, x_1, ..., x_p)' = (z_0, z_1, ..., z_{j+p-1})'$$

and

$$\boldsymbol{\lambda}'_{j-1} = (\beta_0, \beta_1, ..., \beta_{j-1}, \gamma_1, ..., \gamma_p)' = (\lambda_0, \lambda_1, ..., \lambda_{j+p-1}).$$

Now, let us define

$$z_{jk} = \begin{cases} 2y_q - 1, & \text{if } q < j \\ 0, & \text{if } q \geq j \end{cases}, \quad q=1,2,\ldots,j-1.$$

It is evident that $z_{jk} = -1, 0, 1$. Then the likelihood function is

$$L = \prod_{i=1}^{n} \prod_{j=1}^{T} P(y_{ij} \mid y_{i0} y_{i1}, \ldots, y_{ij-1}; x_i) = \prod_{i=1}^{n} \prod_{j=1}^{T} \frac{e^{\theta_j y_{ij}}}{1 + e^{\theta_j}}.$$

The expressions for log likelihood and also the score functions remain same as the previous method.

In the following model,

$$\theta_j = \boldsymbol{\lambda}'_{j-1} \mathbf{z}_{j-1} = \lambda_0 + \lambda_1 z_{j1} + \ldots + \lambda_{j-1} z_{j,j-1} + \lambda_j z_{j,j}.$$

If we define

S_q^+ = number of 1's among the first q outcomes,

S_q^- = number of -1's among the first q outcomes

and $\lambda_q = \lambda^+, \text{ for } z_q = 1,\ \lambda_q = \lambda^-, \text{ for } z_q = -1$

then the model is

$$\theta_j = \boldsymbol{\lambda}'_{j-1} \mathbf{z}_{j-1} = \lambda_0 + \lambda^+ S_{j-1}^+ + \lambda^- S_{j-1}^- + \lambda_j z_j + \ldots + \lambda_{j+p-1} z_{j+p-1}.$$

The equally predictive outcomes, as shown earlier, can be tested as follows:

$$\lambda_1 = \lambda_2 = \ldots = \lambda_q = \lambda \text{ and } S_q = z_1 + \ldots + z_q.$$

Then the model is

$$\theta_j = \lambda_0 + \lambda S_{j-1} + \lambda_j z_j + \ldots + \lambda_{j+p-1} z_{j+p-1}.$$

The same model can be applied under the null hypothesis $\lambda^+ = \lambda^-$.

10.6 APPLICATIONS

In this section, we have used the same HRS data on mobility of elderly population for the period 1992-2004. We have defined two states of the mobility index as 0= no difficulty, 1= difficulty in one or more of the five tasks.

Table 10.1 displays the logistic regression models with previous outcomes as covariates. We observe that the models indicate significant association between previous outcomes and current status even upto the seventh order of Markov chain model. Here the outcomes are the mobility index with two states, no difficulty and some difficulty. To illustrate this example further, let us consider three covariates, age, gender and race (White and Black are two design variables for race categories with other race as the reference category) in addition to previous outcome as a covariate.

Table 10.1. Estimates of Two States Regressive Logistic Model for Mobility Index by Taking Previous Outcomes as Covariates

	Proposed Model			Bonney's Model		
Variables	**β**	**Std. error**	**p-value**	**β**	**Std. error**	**p-value**
Second Order						
Y1	2.604	0.024	0.000	1.185	0.026	0.000
Constant	-1.385	0.015	0.000	-0.284	0.010	0.000
Model χ^2 (p-value)	14502.18	(0.000)		2384.42	(0.000)	
Third Order						
Y1	2.023	0.029	0.000	1.064	0.030	0.000
Y2	1.540	0.030	0.000	0.615	0.030	0.000
Constant	-1.693	0.019	0.000	-0.218	0.011	0.000
Model χ^2 (p-value)	14744.91	(0.000)		3104.93	(0.000)	
Fourth Order						
Y1	1.779	0.034	0.000	0.971	0.032	0.000
Y2	1.269	0.036	0.000	0.642	0.033	0.000
Y3	0.926	0.037	0.000	0.274	0.033	0.000
Constant	-1.788	0.023	0.000	-0.154	0.013	0.000
Model χ^2 (p-value)	12186.37	(0.000)		3326.43	(0.000)	
Fifth Order						
Y1	1.648	0.040	0.000	0.862	0.035	0.000
Y2	1.128	0.043	0.000	0.564	0.037	0.000
Y3	0.810	0.045	0.000	0.357	0.037	0.000
Y4	0.628	0.045	0.000	0.182	0.036	0.000
Constant	-1.800	0.027	0.000	-0.069	0.016	0.000
Model χ^2 (p-value)	9008.75	(0.000)		3071.547	(0.000)	
Sixth Order						
Y1	1.579	0.050	0.000	0.769	0.035	0.000
Y2	1.081	0.054	0.000	0.547	0.037	0.000
Y3	0.751	0.057	0.000	0.355	0.039	0.000
Y4	0.524	0.057	0.000	0.231	0.039	0.000
Y5	0.494	0.057	0.000	0.086	0.038	0.023
Constant	-1.782	0.034	0.000	0.075	0.021	0.000
Model χ^2 (p-value)	5969.25	(0.000)		2898.89	(0.000)	
Seventh Order						
Y1	1.593	0.072	0.000	0.796	0.036	0.000
Y2	1.023	0.077	0.000	0.511	0.039	0.000
Y3	0.715	0.082	0.000	0.357	0.041	0.000
Y4	0.435	0.084	0.000	0.218	0.042	0.000
Y5	0.499	0.084	0.000	0.250	0.042	0.000
Y6	0.380	0.083	0.000	0.190	0.042	0.000
Constant	-1.796	0.049	0.000	0.527	0.041	0.000
Model χ^2 (p-value)	2989.52	(0.000)		2989.52	(0.000)	

Table 10.2 displays from the Bonney's original model that age is consistently positive upto the seventh order model, although shows significant association upto the fourth order

transition. Gender shows negative association with the outcome variable of difficulty in mobility upto the sixth order model. Similarly, Whites show negative association upto the third order and Blacks show positive association upto the fifth order models. However, previous outcomes indicate that even at the seventh order transition probability, the previous outcomes still exert statistically significant positive impacts. The results are slightly different for the selected covariates in the proposed model.

Table 10.2. Estimates of Two States Regressive Logistic Model for Mobility Index by Taking Selected Variables along with Previous Outcomes as Covariate

	Proposed Model			Bonney's Model		
Variables	**β**	**Std. error**	**p-value**	**β**	**Std. error**	**p-value**
Second Order						
Age	0.020	0.003	0.000	0.023	0.002	0.000
Gender	-0.415	0.024	0.000	-0.630	0.020	0.000
White	-0.107	0.064	0.092	-0.161	0.054	0.003
Black	0.131	0.069	0.056	0.198	0.059	0.001
Y1	2.535	0.024	0.000	1.125	0.027	0.000
Constant	-2.365	0.170	0.000	-1.338	0.149	0.000
Model χ^2 (p-value)	14939.08	(0.000)		3684.92	(0.000)	
Third Order						
Age	0.016	0.003	0.000	0.014	0.003	0.000
Gender	-0.285	0.028	0.000	-0.593	0.023	0.000
White	-0.047	0.074	0.527	-0.129	0.061	0.036
Black	0.140	0.080	0.080	0.227	0.066	0.001
Y1	1.991	0.029	0.000	1.034	0.030	0.000
Y2	1.502	0.030	0.000	0.587	0.031	0.000
Constant	-2.540	0.216	0.000	-0.800	0.185	0.000
Model χ^2 (p-value)	14904.62	(0.000)		3984.79	(0.000)	
Fourth Order						
Age	0.019	0.004	0.000	0.010	0.003	0.003
Gender	-0.206	0.032	0.000	-0.554	0.026	0.000
White	-0.007	0.086	0.936	-0.095	0.071	0.176
Black	0.123	0.092	0.182	0.223	0.076	0.003
Y1	1.762	0.034	0.000	0.955	0.032	0.000
Y2	1.250	0.036	0.000	0.618	0.034	0.000
Y3	0.900	0.037	0.000	0.257	0.034	0.000
Constant	-2.931	0.272	0.000	-0.526	0.233	0.024
Model χ^2 (p-value)	12261.34	(0.000)		3889.30	(0.000)	

Table 10.2. Continued

	Proposed Model			Bonney's Model		
Variables	**β**	**Std. error**	**p-value**	**β**	**Std. error**	**p-value**
Fifth Order						
Age	0.019	0.005	0.000	0.004	0.004	0.333
Gender	-0.143	0.038	0.000	-0.502	0.031	0.000
White	-0.006	0.101	0.956	-0.085	0.085	0.315
Black	0.100	0.109	0.359	0.200	0.091	0.029
Y1	1.638	0.041	0.000	0.848	0.035	0.000
Y2	1.120	0.043	0.000	0.550	0.037	0.000
Y3	0.795	0.045	0.000	0.342	0.037	0.000
Y4	0.61	0.045	0.000	0.174	0.037	0.000
Constant	-2.972	0.353	0.000	-0.094	0.303	0.758
Model χ^2 (p-value)	9041.31	(0.000)		3391.15	(0.000)	
Sixth Order						
Age	0.019	0.007	0.006	0.002	0.006	0.773
Gender	-0.153	0.047	0.001	-0.458	0.040	0.000
White	0.037	0.126	0.770	-0.053	0.109	0.626
Black	0.079	0.136	0.561	0.150	0.117	0.202
Y1	1.571	0.050	0.000	0.770	0.035	0.000
Y2	1.075	0.054	0.000	0.531	0.038	0.000
Y3	0.740	0.057	0.000	0.341	0.039	0.000
Y4	0.511	0.058	0.000	0.218	0.039	0.000
Y5	0.487	0.057	0.000	0.080	0.038	0.035
Constant	-3.012	0.476	0.000	0.178	0.417	0.669
Model χ^2 (p-value)	5988.45	(0.000)		3050.08	(0.000)	
Seventh Order						
Age	0.018	0.010	0.086	0.018	0.010	0.086
Gender	-0.111	0.067	0.097	-0.111	0.067	0.097
White	-0.037	0.180	0.838	-0.037	0.180	0.838
Black	-0.077	0.195	0.691	-0.077	0.195	0.691
Y1	1.586	0.072	0.000	0.793	0.036	0.000
Y2	1.023	0.077	0.000	0.511	0.039	0.000
Y3	0.707	0.083	0.000	0.354	0.041	0.000
Y4	0.430	0.084	0.000	0.215	0.042	0.000
Y5	0.49	0.084	0.000	0.25	0.042	0.000
Y6	0.38	0.083	0.000	0.19	0.042	0.000
Constant	-2.897	0.720	0.0001	-0.588	0.718	0.413
Model χ^2 (p-value)	2995.46	(0.000)		2995.46	(0.000)	

Table 10.1A and Table 10.2A show the test for equally predictive prior observations for conditional probabilities of Y_j for given $Y_1,...,Y_{j-1}$. We observe that the estimates of the parameters for S^- are negative but the estimates for S^+ are positive. Ideally, under the equally predictive observations, the conditional probability of Y_j for given $Y_1,...,Y_{j-1}$ is expected to increase the logit by the same amount for S^+ as is expected to decrease for S^-. Table 10.1A shows the models for mobility index to test for equally predictive observations. In all the models from second to sixth order, the positive effects are higher than that of the negative effects which is indicative of the fact that the observations are not equally predictive rather there is positive impact of the event in the past to the current event. Similar results are obtained if we consider the equally predictive observations in the presence of other covariates. It is also noteworthy from Table 10.1A and Table 10.2A that after each order the amount of impact on the current event is reduced.

Table 10.1a. Estimates of Two States Regressive Logistic Model for Mobility Index by Taking Previous Outcome as Covariates for Equal Prediction

	Bonney's Model			
Variables	**β**	**Std. error**	**p-value**	**OR**
Second Order				
S^+	1.437	.048	.000	4.207
S^-	-1.051	.033	.000	.350
Constant	-.308	.011	.000	.735
Model χ^2 (p-value)	2425.75	(0.000)		
Third Order				
S^+	1.094	.031	.000	2.985
S^-	-.729	.020	.000	.482
Constant	-.259	.012	.000	.772
Model χ^2 (p-value)	3137.64	(0.000)		
Fourth Order				
S^+	.818	.022	.000	2.266
S^-	-.553	.015	.000	.575
Constant	-.202	.014	.000	.817
Model χ^2 (p-value)	3254.57	(0.000)		
Fifth Order				
S^+	.647	.019	.000	1.910
S^-	-.434	.012	.000	.648
Constant	-.131	.018	.000	.877
Model χ^2 (p-value)	2988.69	(0.000)		
Sixth Order				
S^+	.569	.017	.000	1.766
S^-	-.334	.010	.000	.716
Constant	-.072	.025	.004	.931
Model χ^2 (p-value)	2834.92	(0.000)		

Table 10.2a. Estimates of Two States Regressive Logistic Model for Mobility Index by Taking All Covariates along with Previous Outcomes as Covariates for Equal Prediction

Bonney's Model				
Variables	**β**	**Std. err**	**p-value**	**OR**
Second Order				
Age	.034	.002	.000	1.035
Gender	-.628	.020	.000	.533
White	-.163	.055	.003	.849
Black	.194	.059	.001	1.214
S^+	1.577	.051	.000	4.839
S^-	-.848	.036	.000	.428
Constant	-2.087	.165	.000	.124
Model χ^2 (p-value)	3804.86	(0.000)		
Third Order				
Age	.028	.003	.000	1.028
Gender	-.591	.023	.000	.554
White	-.131	.061	.033	.877
Black	.221	.066	.001	1.247
S^+	1.119	.031	.000	3.061
S^-	-.653	.022	.000	.520
Constant	-1.706	.201	.000	.182
Model χ^2 (p-value)	4066.65	(0.000)		
Fourth Order				
Age	.023	.004	.000	1.024
Gender	-.551	.026	.000	.576
White	-.091	.071	.196	.913
Black	.223	.076	.003	1.250
S^+	.823	.023	.000	2.276
S^-	-.512	.016	.000	.599
Constant	-1.439	.252	.000	.237
Model χ^2 (p-value)	3842.07	(0.000)		
Fifth Order				
Age	.017	.005	.000	1.017
Gender	-.496	.031	.000	.609
White	-.080	.084	.341	.923
Black	.200	.091	.028	1.222
S^+	.640	.019	.000	1.896
S^-	-.412	.013	.000	.662
Constant	-.998	.324	.002	.369
Model χ^2 (p-value)	3312.48	(0.000)		

Table 10.2a. Continued

Variables	β	Std. err	p-value	OR
		Bonney's Model		
Sixth Order				
Age	.018	.006	.004	1.018
Gender	-.438	.040	.000	.645
White	-.025	.108	.818	.975
Black	.164	.117	.160	1.178
S^+	.562	.017	.000	1.753
S^-	-.320	.011	.000	.726
Constant	-1.089	.433	.012	.337
Model χ^2 (p-value)	2981.23			
	3	(0.000)		

10.7 Crowley-Jones Method and Simple Alternatives

In this section, we examine the performance of several test procedures for both whether a Markov model holds as well as the significance of parameters associated with transition probabilities. Jones and Crowley (1992) proposed nonparametric tests of the Markov model for survival data but the approach can be applied to one way transition only. Jones and Crowley (1992) assumed that the individuals enter the waiting time state s_0 at time t=0. After entering into state s_0, some of them can either move to intermediate state s_1 or some can remain in state s_0.Then from both s_0 and s_1, the individuals may proceed to failure state 0 or competing state C. Jones and Crowley proposed a test procedure for testing whether transition rate from state s_1 to the failure state 0 depends on the waiting time W in the waiting state s_0. This procedure is applicable for one way transition only.

We addressed two problems in this section: (i) to generalize Jones and Crowley (1992) approach for taking into account transitions and reverse transitions which emerge most frequently from longitudinal data, and (ii) to propose an alternative test procedure which is more flexible for addressing both estimation of parameters in order to examine the influence of risk factors on transition probabilities as well as to examine whether the Markov model holds at all.

10.7.1 Estimation Based on Two State Markov Chain

The transition matrix π of a two state Markov Chain is defined as

$$\pi = \begin{pmatrix} \pi_{00} & \pi_{01} \\ \pi_{10} & \pi_{11} \end{pmatrix},$$

where $\sum_{m=0}^{1} \pi_{sm} = 1$, s=0,1.

Let $Y_{ij} = 1$ if an event occurs to the individual at the j-th follow-up and $Y_{ij} = 0$, otherwise. There are alternative approaches to link transition probabilities with corresponding covariates. Two such models are described below.

Model I: Let X_{ij} be the vector of p covariates for the i-th individual (i=1,2,…,n) at the j-th follow-up (j=1,2,…,J) where $X_{ij}' = \left(X_{ij1}, X_{ij2}, ..., X_{ijp}\right)$, $\beta' = \left(\beta_1, \beta_2, ..., \beta_p\right)$ and let Y_{ij} be the outcome variable for the i-th individual at the j-th follow-up. Then we can express the model as follows assuming first order Markov chain:

$$\text{logit } P(Y_{ij} = 1 | \; Y_{ij-1}, Y_{ij-2}, ..., Y_{i1}) = X_{ij}'\beta \; + \alpha Y_{ij-1}, \qquad (10.9)$$

where, we can show, following Markov model

$$\pi_{01ij} = P(Y_{ij} = 1 | Y_{ij-1} = 0) = \frac{\exp(X_{ij}'\beta)}{1 + \exp(X_{ij}'\beta)}$$

and

$$\pi_{10ij} = P(Y_{ij} = 1 | Y_{ij-1} = 1) = \frac{\exp(X_{ij}'\beta + \alpha)}{1 + \exp(X_{ij}'\beta + \alpha)}.$$

Model II: This model was proposed by Muenz and Rubinstein (1985). This model considers the following covariate dependent Markov model:

$$\text{logit } P(Y_{ij} = m | \; Y_{ij-1} = s) = X_{ij}'\beta_s, \; s \neq m, \; (s,m=0,1). \qquad (10.10)$$

Then we can define the following conditional probabilities based on Markov property:

$$\pi_{01ij} = P(Y_{ij} = 1 | \; Y_{ij-1} = 0) = \frac{\exp(X_{ij}'\beta_{01})}{1 + \exp(X_{ij}'\beta_{01})},$$

and

$$\pi_{10ij} = P(Y_{ij} = 1 | \; Y_{ij-1} = 1) = \frac{\exp(X_{ij}'\beta_{10})}{1 + \exp(X_{ij}'\beta_{10})}.$$

Between the alternative models (10.9) and (10.10), the later (Model II) appears to be more flexible for addressing the real life problems.

10.7.2 A Test for Significance of Waiting Time

A test for waiting time is proposed here on the basis of the Jones and Crowley (1992) method. Let us consider a sample of size n. The i-th individual is followed n_i times. The transition probabilities can be displayed as follows

Y_{j-1}		Y_j
	0	1
0	π_{00}	π_{01}.
1	π_{10}	π_{10}

Jones and Crowley (1992) suggested a nonparametric method of testing for the Markov model but the approach can be employed only for one-way transitions. In this section, we extend the Jones and Crowley test for both transitions and reverse transitions as we observe from the above transition probability matrix. If we consider the transition $0 \to 1$, and reverse transition $1 \to 0$ as two distinct transition types, then for transition type $0 \to 1$ we want to examine whether $\lambda_0(t,w) = \lambda_0(t)$.

Similarly, we can define for transition type $1 \to 0$, $\lambda_1(t,w) = \lambda_1(t)$.

In other words, for s-th (transition from s, s=0,1) transition type, we want to test for $\lambda_s(t,w) = \lambda_s(t)$, s=0,1.

Let us suppose that there are k_0 distinct transition times for the transition of the type $0 \to 1$ and the ordered transition times from state 0 are $t_1^{(0)} < t_2^{(0)} < ... < t_{k_0}^{(0)}$.

Similarly, suppose that there are k_1 distinct transition times for the transition of the type $1 \to 0$ and the ordered transition times from state 1 are $t_1^{(1)} < t_2^{(1)} < ... < t_{k_1}^{(1)}$.

Let $t_1^{(s)} < t_2^{(s)} < ... < t_{k_s}^{(s)}$ (s=0,1) represent the ordered transition times from state s (s=0,1):

Let

n_s = total number of individuals in state s (s=0,1) at the beginning of the study,

$$\delta_j^{(s)}(t) = \begin{cases} 1, & \text{if the jth individual is in state s (s=0,1) at time t,} \\ 0, & \text{otherwise;} \end{cases}$$

$$J_j^{(s)}(t) = \begin{cases} 1, & \text{if the jth individual is in state s (s=0,1) and is observed to fail at time t,} \\ 0, & \text{otherwise.} \end{cases}$$

Then we can also define

$n^{(s)}(t) = \sum_{j=1}^{n_s} \delta_j^{(s)}(t) =$ number of individuals at risk of making transition from state s at time t,

$d^{(s)}(t) = \sum_{j=1}^{n_s} J_j^{(s)}(t) =$ number of individuals that make transitions from state s at time t,

$W_j^{(s)} =$ waiting time for the jth individual in state s (s=0,1).

The model for waiting time is

$$\lambda_s(t, W) = \lambda_s(t)\exp\left[W^{(s)}\beta_{sw}\right], \quad s=0,1,$$

where β_{sw} s are the coefficient of waiting time for a transition from state s (s=0,1). To test for the waiting time, our null hypothesis is

$$H_0 : \begin{pmatrix} \beta_{oW} \\ \beta_{1W} \end{pmatrix} = \begin{pmatrix} 0 \\ 0 \end{pmatrix}.$$

Generalizing the test statistic suggested by Jones and Crowley (1992) for both transitions $(0 \to 1)$ and reverse transitions $(1 \to 0)$, we obtain

$$T(q,z) = \sum_{s=0}^{1} \sum_{i=1}^{k_s} q^{(s)}(t_i) \sum_{j=1}^{n_s} J_j^{(s)}(t_i)\left[Z_j^{(s)}(t_i) - \overline{Z}^{(s)}(t_i)\right],$$

and

$$q^{(s)}(t_i) = \text{weight function at time } t_i = \begin{cases} n^s(t_i)/n_s, & \text{case I} \\ 1, & \text{case II} \end{cases},$$

$Z_j^{(s)}(t_i) =$ function of waiting time for the jth individual in state s (s=0,1), $W_j^{(s)}$,

$$\overline{Z}^{(s)}(t_i) = \sum_{j=1}^{n_s} \frac{\delta_j^{(s)}(t_i) Z_j^{(s)}(t_i)}{n^{(s)}(t)} = \text{average label of those at risk at time } t_i.$$

We may consider the most obvious choice for the function of waiting time,

$$Z_j^{(s)}(t_i) = W_j^{(s)}.$$

Assuming independence of waiting times in states 0 and 1, the variance can be obtained as follows:

$$V(q,t) = \sum_{s=0}^{1} \sum_{i=1}^{k_s} \left[q^{(s)}(t_i)\right]^2 d^{*(s)}(t_i)\sigma_{zz}^{(s)}(t_i),$$

where

$$d^{*(s)}(t_i) = \frac{d^{(s)}(t_i)\left[n^{(s)}(t_i) - d^{(s)}(t_i)\right]}{n^{(s)}(t_i) - 1},$$

and

$$\sigma_{zz}^{(s)}(t_i) = \frac{1}{n^{(s)}(t_i)} \sum_{j=1}^{n_s} \delta_j^{(s)}(t_i)\left[Z_j^{(s)}(t_i) - \bar{Z}^{(s)}(t_i)\right]^2$$

is the sample variance of the labels of those at risk at time t_i. If there are no transition times with ties then $d^{*(s)} \equiv 1$. The test statistic is

$$T/\sqrt{V} \sim N(0,1).$$

There are two strong limitations of the use of the weights undertaken here: (i) independence between entry time and survival time is questionable (Case I), and (ii) $n^{(s)}(t)$ rises quickly at first, is fairly constant for a while, then decreases very gradually (Case II).

10.8 Alternative Test Procedures

In this section an alternative test procedure is proposed to take into account other covariates, in addition to the waiting time, for a Markov model comprising of transitions and reverse transitions. Here, two test procedures are proposed as extensions of Models I and II. It is worth noting that in the existing literature there is no suitable technique for testing the adequacy of the Markov model as well as for testing the covariate dependence of the Markov models. The proposed test procedures address both the adequacy of the Markov model as well as their covariate dependence simultaneously.

Alternative Test Procedure Based on Model I

An alternative test procedure is presented here based on Model I. In Model I, only outcome of the previous follow-up needs to be incorporated as covariate and we can test for the corresponding parameter to test for the Markov model. Instead of only previous outcome, we can incorporate all the previous outcomes at different follow-ups to make the test procedure more flexible to investigate for the adequacy of Markov model as well as likely deviations from the first order Markov model.

Let X_{ij} be the vector of p covariates for the i-th individual (i=1,2,…,n) at the j-th follow-up (j=1,2,…,J) where $X_{ij}^{/} = \left(1, X_{ij1}, X_{ij2}, ..., X_{ijp}\right), \beta^{/} = \left(\beta_0, \beta_1, \beta_2, ..., \beta_p\right)$, e^{α} = ratio of odds of occurrence of event at the j-th follow-up among individuals who did or did not experience the event at the (j-1)th follow-up and Y_{ij} be the outcome variable for the i-th (i=1,2,…, n_s ; s=0,1) individual at the j-th (j=1,2,…, r_i) follow-up. Another interpretation of α is that it signifies the Markov relationship during consecutive follow-up visits. Then we can express the model as follows:

$$\text{logit } P(Y_{ij}=1\,|H_{ij})=X_{ij}^{\prime}\beta+\alpha Y_{ij-1},$$

where we can show following Markov model

$$\pi_{01ij} = P(Y_{ij} = 1|\ Y_{ij-1} = 0) = \frac{\exp(X_{ij}^{\prime}\beta)}{1+\exp(X_{ij}^{\prime}\beta)},$$

and

$$\pi_{10ij} = P(Y_{ij} = 0|\ Y_{ij-1} = 1) = \frac{\exp(X_{ij}^{\prime}\beta\ +\alpha)}{1+\exp(X_{ij}^{\prime}\beta\ +\alpha)}.$$

The likelihood function is

$$L=\prod_{s=0}^{1}\prod_{m=0}^{1}\prod_{i=1}^{n_s}\prod_{j=1}^{r_i}\left[\pi_{smij}\right]^{\delta_{smij}}.$$

We can obtain the log likelihood function from the above expression and the estimates of the parameters can be obtained by setting first derivatives of the log likelihood function with respect to the parameters equal to zero. Then the tests can be performed in two stages

I. The likelihood ratio test can be employed to test for the significance of the overall model at the first stage.

II. The Wald test can be applied to test for the significance of the parameter(s) corresponding to the previous outcomes as shown below:

$$W=\hat{\alpha}\,/\,se(\hat{\alpha}).$$

This procedure can be extended in a straightforward manner for a j-th order Markov model as

$$\text{logit } P(Y_{ij}|\ Y_{ij-1},Y_{ij-2},\dots,Y_{i1})=X_{ij}^{\prime}\beta\ +\alpha_1 Y_{ij-1}+\alpha_2 Y_{ij-2}+\dots+\alpha_{j-1}Y_{i1},$$

In this case, there will be j-1 Wald tests to be performed.

Alternative Test Procedure Based on Model II

The covariate dependent Markov models can be expressed as shown below:

$$\pi_{01ij} = P(Y_{ij} = 1|\ Y_{ij-1} = 0) = \frac{\exp(X_{ij}^{\prime}\beta_{01})}{1+\exp(X_{ij}^{\prime}\beta_{01})},$$

and

$$\pi_{10ij} = P(Y_{ij} = 0 \mid Y_{ij-1} = 1) = \frac{\exp(X_{ij}^{/}\beta_{10})}{1+\exp(X_{ij}^{/}\beta_{10})}.$$

As defined in Section 10.7.1 for Model II, X_{ij} is the vector of p covariates for the i-th individual (i=1,2,…,n) at the j-th follow-up (j=1,2,…,J) where $X_{ij}^{/} = \left(X_{ij1}, X_{ij2}, ..., X_{ijp}\right)$; and Y_{ij} is the outcome variable for the i-th individual at the j-th follow-up.

To test for the role of waiting time, W_{ij}, in states 0 and 1; the transition probabilities can be reformulated as follows:

$$\pi_{01ij} = P(Y_{ij} = 1 \mid Y_{ij-1} = 0, W_{ij}) = \frac{\exp(X_{ij}^{/}\beta_{01} + W_{ij}\beta_{01w})}{1+\exp(X_{ij}^{/}\beta_{01} + W_{ij}\beta_{01w})},$$

and

$$\pi_{10ij} = P(Y_{ij} = 0 \mid Y_{ij-1} = 1, W_{ij}) = \frac{\exp(X_{ij}^{/}\beta_{10} + W_{ij}\beta_{10w})}{1+\exp(X_{ij}^{/}\beta_{10} + W_{ij}\beta_{10w})}.$$

The estimates for the parameters can be obtained from the following equations

$$\frac{\partial \ln L}{\partial \beta^{*}_{smq}} = \sum_{i=1}^{n_s}\sum_{j=1}^{r_i} X^{*}_{ijq}\left[\delta_{smij} - (\delta_{s0ij} + \delta_{s1ij})\frac{\exp(X_{ij}^{/*}\beta^{*}_{sm})}{1+\exp(X_{ij}^{/*}\beta^{*}_{sm})}\right] = 0$$

s,m=0,1 (s$\neq$ m), q=0,1,2,…,p,w,

$$\frac{\partial^2 \ln L}{\partial \beta^{*}_{smq}\partial \beta^{*}_{sml}} = -\sum_{i=1}^{n_s}\sum_{j=1}^{r_i} X^{*}_{ijq}\, X^{*}_{ijl}\left[(\delta_{smij} + \delta_{smil})\left(\frac{\exp(X_{ij}^{/*}\beta^{*}_{sm})}{1+\exp(X_{ij}^{/*}\beta^{*}_{sm})}\right)\left(\frac{\exp(X_{il}^{/*}\beta^{*}_{sm})}{1+\exp(X_{il}^{/*}\beta^{*}_{sm})}\right)\right]$$

where n_s = number of entries in state s (s=0,1), r_i = number of follow-ups observed for individual i (I=1,2,…, n_s ,s=0,1) $X_{ij}^{/*} = \left(X_{ij1}, X_{ij2}, ..., X_{ijp}, W_{ij}\right)$,

$\beta_{sm}^{*\,/} = (\beta_{sm1}, \beta_{sm2}, ..., \beta_{smp}, \beta_{smw})$ and δ_{smij}=1 if the i-th individual makes a transition of the type $s \rightarrow m$ (s=0,1; m=0,1) at the j-th follow-up, δ_{smij}=0, otherwise.

The tests can be performed in two stages for testing the null hypothesis for the overall model and then for the waiting times as follows:

I. The likelihood ratio test can be employed to test for the significance of the overall model at the first stage.

II. The Wald test can be applied to test for the significance of the waiting time on the transition probabilities as follows:

$$W^{(s)} = \hat{\beta}_{smw} / se(\hat{\beta}_{smw}),\ s \neq m.$$

This procedure can also be generalized for higher order Markov models as indicated for Model I.

10.9 APPLICATION

We have used data from the longitudinal study from the survey on Maternal Morbidity in Bangladesh conducted by the Bangladesh Institute for Research for Promotion of Essential and Reproductive Health Technologies (BIRPERHT) during November 1992 to December 1993. All the subjects selected in this study are comprised of pregnant women with less than 6 months duration. They were followed roughly at an interval of one month throughout the pregnancy. For the purpose of this study, we have selected 993 pregnant women, with at least one antenatal follow-up. Table 10.3 shows the number of respondents at different follow-up visits during antenatal period. At the first follow-up 992 respondents were recorded (out of 993 respondents, 1 was missing at the first follow-up but reported subsequently). The number dropped to 917 at the second follow-up and the rate of drop-out increased sharply at subsequent follow-ups. The number of respondents observed at the third and the fourth follow-ups were 771 and 594, respectively. The number decreased for the higher follow-ups due to late entry of a substantial number of pregnant women into the study before delivery.

Table 10.3. Number of Respondents at Different Follow-ups during Antenatal Period

Follow-upNumber	Frequency
1	992
2	917
3	771
4	594
5	370
6	148

The following pregnancy complications are considered under the complications in this study: hemorrhage, edema, excessive vomiting, fits/convulsion. If one or more of these complications occurred to the respondents, they were considered as having complications.

The explanatory variables are: pregnancies prior to the index pregnancy (yes, no), education of respondent (no schooling, some schooling), age at marriage (less than 15 years, 15 years or more), involved with gainful employment (no, yes), index pregnancy was wanted or not (no, yes).

The application of the extended Jones-Crowley test for transition and reverse transition indicate that the null hypothesis of no role of waiting time might be rejected for both the assumptions for weights (Case I and Case II, as described in 10.7.2) which are

$$q^{(s)}(t_i) = \text{ weight function at time } t_i = \begin{cases} n^s(t_i)/n_s, & \text{case I} \\ 1, & \text{case II} \end{cases}.$$

The test statistic value for Case I is 115.12 and for Case II is 179.55.

Table 10.4 displays the estimates of parameters based on Model I. It is evident from the tests that among the selected covariates, wanted pregnancy is seemed to have significant association with transitions between no complications to complications or complications to no complications during pregnancy. In addition, it is important that, there is evidence of statistically significant association between consecutive outcomes. This proves the usefulness of fitting a first order Markov model.

Table 10.4. Estimates of the Parameters of Model I

Variables	Estimates	Std. error	t-value	p-value
Constant	-1.010	0.107	-9.399	0.000
Wanted pregnancy (Yes=1)	-0.402	0.094	-4.284	0.000
Gainful employment (Yes=1)	-0.149	0.100	-1.493	0.135
Age at marriage (< 15 = 1)	0.024	0.096	0.251	0.802
Education (Yes=1)	0.121	0.092	1.311	0.190
Previous pregnancies (Yes=1)	-0.044	0.094	-0.468	0.640
α	0.414	0.145	2.858	0.004
Likelihood Ratio Test		925.408; d.f. = 7	(p=0.00000)	

Table 10.5 displays the estimates based on logistic regression models for each follow-up, taking previous outcome as a covariate. We observe that different covariates display significant association with transition probabilities at different follow-ups, but for all the follow-ups the previous outcome show significant association indicating the adequacy of the first order Markov model.

Table 10.6 shows the extended Model I, which incorporate all the previous outcomes. The results show that wanted pregnancy and age at marriage are significantly associated at follow-up 1, education and status at the first follow-up are associated with follow-up 2, status at follow-ups 1 and 2 are significantly associated with the outcome at follow-up 3, gainful employment, and status at follow-ups 1,2 and 3 are associated with the outcome at follow-up 4, status at follow-ups 3 and 4 are associated with the outcome at follow-up 5, and gainful employment and status at follow-up 5 are associated with follow-up 6. These results indicate that the assumption of first order Markov model may not be adequate in strict sense as observed from the findings based on Model I. However, Tables 3 and 4 are simplified analysis of data to examine whether there is likely evidence that may contradict the results based on Model I. This requires further investigation.

The alternative test procedure based on Model II is presented in Table 10.8. The number of transitions and reverse transitions are displayed in Table 10.7. Table 10.8 shows that the overall likelihood ratio test confirms significance of the proposed model. For transitions, wanted pregnancy and waiting time appear to be significant but for reverse transitions wanted pregnancy, gainful employment and education are statistically significant. These results imply that the null hypothesis is accepted for reverse transition but rejected for transition. In other words, the Markov assumption can be considered for reverse transition, but the transition appears to favor a semi Markov model.

Table 10.5. The Estimates of the Logistic Regression Model Parameters at Different Follow-ups Incorporating Previous Outcome as a Covariate

Model and Variables	Coefficients	Standard error	Wald	p-value	Odds Ratio
Follow-up 1					
Wanted Pregnancy	-0.606	0.143	17.952	0.000	0.545
Gainful employment	-0.224	0.147	2.306	0.129	0.799
Age at marriage	-0.287	0.144	3.979	0.046	0.750
Education	-0.164	0.137	1.417	0.234	0.849
Number of pregnancies	-0.191	0.139	1.895	0.169	0.826
Constant	0.431	0.162	7.098	0.008	1.539
Model Chi-Square (p-value)			28.165 (0.000)		
Follow-up 2					
Wanted Pregnancy	0.180	0.179	1.011	0.315	1.198
Gainful employment	-0.082	0.183	0.204	0.652	0.921
Age at marriage	-0.085	0.181	0.218	0.641	0.919
Education	-0.550	0.173	10.056	0.002	0.577
Number of pregnancies	-0.052	0.173	0.089	0.765	0.950
Status at 1st follow up	2.533	0.179	201.288	0.000	12.585
Constant	-1.922	0.240	64.264	0.000	0.146
Model Chi-Square (p-value)			271.536 (0.000)		
Follow-up 3					
Wanted Pregnancy	0.197	0.207	0.900	0.343	1.218
Gainful employment	0.199	0.209	0.908	0.341	1.221
Age at marriage	-0.037	0.207	0.032	0.858	0.964
Education	-0.155	0.199	0.602	0.438	0.857
Number of pregnancies	-0.002	0.201	0.000	0.991	0.998
Status at 2nd follow up	2.771	0.196	199.996	0.000	15.975
Constant	-2.014	0.263	58.649	0.000	0.133
Model Chi-Square (p-value)			249.061 (0.000)		
Follow-up 4					
Wanted Pregnancy	0.129	0.226	0.327	0.568	1.138
Gainful employment	0.671	0.226	8.839	0.003	1.957
Age at marriage	-0.066	0.232	0.082	0.774	0.936
Education	-0.338	0.221	2.342	0.126	0.713
Number of pregnancies	0.420	0.219	3.674	0.055	1.522
Status at 3rd follow up	2.089	0.214	95.033	0.000	8.077
Constant	-2.009	0.290	48.043	0.000	0.134
Model Chi-Square (p-value)			122.700 (0.000)		

Table 10.5. Continued

Model and Variables	Coefficients	Standard error	Wald	p-value	Odds Ratio
Follow-up 5					
Wanted Pregnancy	-0.871	0.260	11.224	0.001	0.419
Gainful employment	-0.289	0.274	1.112	0.292	0.749
Age at marriage	0.155	0.262	0.348	0.555	1.167
Education	-0.335	0.258	1.690	0.194	0.715
Number of pregnancies	-0.129	0.264	0.238	0.626	0.879
Status at 4th follow up	1.884	0.272	47.956	0.000	6.581
Constant	-0.497	0.292	2.891	0.089	0.608
Model Chi-Square (p-value)			66.966 (0.000)		
Follow-up 6					
Wanted Pregnancy	-0.048	0.436	0.012	0.913	0.953
Gainful employment	1.018	0.441	5.332	0.021	2.766
Age at marriage	-0.453	0.459	0.976	0.323	0.636
Education	-0.008	0.423	0.000	0.985	0.992
Number of pregnancies	0.533	0.444	1.438	0.230	1.704
Status at 5th follow up	1.892	0.442	18.347	0.000	6.634
Constant	-1.953	0.557	12.295	0.000	0.142
Model Chi-Square (p-value)			28.981 (0.000)		

Table 10.6. Estimates Based on the Proposed Extension of Model I

Model and Variables	Coefficients	Standard error	Wald	p-value	Odds Ratio
Follow-up 1					
Wanted Pregnancy	-0.606	0.143	17.952	0.000	0.545
Gainful employment	-0.224	0.147	2.306	0.129	0.799
Age at marriage	-0.287	0.144	3.979	0.046	0.750
Education	-0.164	0.137	1.417	0.234	0.849
Number of pregnancies	-0.191	0.139	1.895	0.169	0.826
Constant	0.431	0.162	7.098	0.008	1.539
Model Chi-Square (p-value)			28.165 (0.000)		
Follow-up 2					
Wanted Pregnancy	0.180	0.179	1.011	0.315	1.198
Gainful employment	-0.082	0.183	0.204	0.652	0.921
Age at marriage	-0.085	0.181	0.218	0.641	0.919
Education	-0.550	0.173	10.056	0.002	0.577
Number of pregnancies	-0.052	0.173	0.089	0.765	0.950
Status at 1st follow up	2.533	0.179	201.288	0.000	12.585
Constant	-1.922	0.240	64.264	0.000	0.146
Model Chi-Square (p-value)			271.536 (0.000)		
Follow-up 3					
Wanted Pregnancy	0.301	0.211	2.033	0.154	1.351
Gainful employment	0.230	0.212	1.177	0.278	1.259
Age at marriage	0.005	0.210	0.001	0.982	1.005
Education	-0.181	0.202	0.804	0.370	0.834

Table 10.6. (Continued)

Model and Variables	Coefficients	Standard error	Wald	p-value	Odds Ratio
Number of pregnancies	0.027	0.203	0.018	0.894	1.027
Status at 1st follow up	0.963	0.213	20.446	0.000	2.620
Status at 2nd follow up	2.345	0.212	121.990	0.000	10.435
Constant	-2.413	0.285	71.455	0.000	0.090
Model Chi-Square (p-value)			268.347 (0.000)		
Follow-up 4					
Wanted Pregnancy	0.210	0.235	0.796	0.372	1.233
Gainful employment	0.728	0.237	9.479	0.002	2.072
Age at marriage	-0.089	0.240	0.136	0.712	0.915
Education	-0.222	0.231	0.925	0.336	0.801
Number of pregnancies	0.427	0.228	3.522	0.061	1.533
Status at 1st follow up	0.566	0.249	5.172	0.023	1.761
Status at 2nd follow up	1.066	0.277	14.804	0.000	2.903
Status at 3rd follow up	1.281	0.258	24.681	0.000	3.600
Constant	-2.456	0.323	57.705	0.000	0.086
Model Chi-Square (p-value)			148.013 (0.000)		
Follow-up 5					
Wanted Pregnancy	-0.928	0.298	9.656	0.002	0.396
Gainful employment	-0.316	0.318	0.989	0.320	0.729
Age at marriage	0.245	0.303	0.654	0.419	1.278
Education	-0.113	0.295	0.147	0.701	0.893
Number of pregnancies	0.139	0.298	0.218	0.641	1.149
Status at 1st follow up	0.536	0.321	2.787	0.095	1.709
Status at 2nd follow up	0.626	0.377	2.761	0.097	1.871
Status at 3rd follow up	1.358	0.350	15.024	0.000	3.888
Status at 4th follow up	1.258	0.322	15.278	0.000	3.518
Constant	-1.393	0.371	14.070	0.000	0.248
Model Chi-Square (p-value)			113.846 (0.000)		
Follow-up 6					
Wanted Pregnancy	-0.341	0.474	0.516	0.473	0.711
Gainful employment	1.196	0.499	5.744	0.017	3.307
Age at marriage	-0.703	0.504	1.949	0.163	0.495
Education	0.027	0.465	0.003	0.954	1.027
Number of pregnancies	0.380	0.500	0.578	0.447	1.462
Status at 1st follow up	0.135	0.566	0.057	0.812	1.144
Status at 2nd follow up	0.672	0.655	1.051	0.305	1.958
Status at 3rd follow up	-0.564	0.633	0.794	0.373	0.569
Status at 4th follow up	0.978	0.522	3.509	0.061	2.658
Status at 5th follow up	1.771	0.572	9.601	0.002	5.879
Constant	-1.963	0.620	10.036	0.002	0.140
Model Chi-Square (p-value)			36.133 (0.000)		

Table 10.7. Number of Transitions in Pregnancy Complications

Transitions	**→0**	**→1**
First Order		
0→	1577	277
1→	366	614

Table 10.8. Estimates of the Parameters of Covariate Dependent Markov Models for Analyzing Pregnancy Complications

Variables	Estimates	Std. error	t-value	p-value
	FIRST ORDER			
0→ 1				
Constant	-2.349	0.336	-6.987	0.000
Wanted pregnancy (Yes=1)	-0.393	0.137	-2.869	0.004
Gainful employment (Yes=1)	0.044	0.143	0.310	0.756
Age at marriage (< 15 = 1)	-0.011	0.139	-0.079	0.937
Education (Yes=1)	-0.053	0.134	-0.400	0.689
Previous pregnancies (Yes=1)	0.061	0.135	0.449	0.654
Waiting time	0.150	0.050	2.975	0.003
1→ 0				
Constant	-0.397	0.302	-1.313	0.189
Wanted pregnancy (Yes=1)	-0.300	0.139	-2.151	0.031
Gainful employment (Yes=1)	-0.331	0.148	-2.233	0.026
Age at marriage (< 15 = 1)	0.150	0.147	1.020	0.308
Education (Yes=1)	0.561	0.140	4.001	0.000
Previous pregnancies (Yes=1)	-0.096	0.140	-0.684	0.494
Waiting time	-0.012	0.046	-0.256	0.798
Likelihood Ratio Test	1119.16; d.f. = 14 (p=0.00000)			
LR Test for the Waiting Time	9.09; d.f. = 1 (p=0.002567)			

10.10 SUMMARY

This chapter provides some important test procedures for covariate dependent first or higher order models. The likelihood ratio based tests for models with two or more states are summarized in Section 10.2 and the Wald test for significance of single parameter is displayed in Section 10.3. Test for order is proposed in Section 10.4. Regressive logistic

$$n = \begin{pmatrix} n_{000} & n_{001} \\ n_{010} & n_{011} \\ n_{100} & n_{101} \\ n_{110} & n_{111} \end{pmatrix}, \ \pi = \begin{pmatrix} \pi_{000} & \pi_{001} \\ \pi_{010} & \pi_{011} \\ \pi_{100} & \pi_{101} \\ \pi_{110} & \pi_{111} \end{pmatrix}.$$

Here a 2x2x2 table or two 2x2 tables are expressed as a 4x2 table for convenience.

Let

$\pi_{ij+} = \sum_k \pi_{ijk}$, $\pi_{i+k} = \sum_j \pi_{ijk}$, $\pi_{+jk} = \sum_i \pi_{ijk}$, $\pi_{i++} = \sum_j \sum_k \pi_{ijk}$, $\pi_{+j+} = \sum_i \sum_k \pi_{ijk}$, and $\pi_{++k} = \sum_i \sum_j \pi_{ijk}$.

The three variables are mutually independent when

$$\pi_{ijk} = \pi_{i++}\pi_{+j+}\pi_{++k}, \text{ for all i,j and k.}$$

For expected frequencies, mutual independence has loglinear form:

$$\log \mu_{ijk} = \lambda + \lambda_i^{Y_{t-2}} + \lambda_j^{Y_{t-1}} + \lambda_k^{Y_t}.$$

Variable Y_j is jointly independent of Y_{j-2} and Y_{j-1} when

$$\pi_{ijk} = \pi_{ij+}\pi_{++k} \quad \text{for all i,j,k.}$$

The loglinear model is

$$\log \mu_{ijk} = \lambda + \lambda_i^{Y_{t-2}} + \lambda_j^{Y_{t-1}} + \lambda_k^{Y_t} + \lambda_{ij}^{Y_{t-2}Y_{t-1}}.$$

Similarly, we can also show that if Y_{t-2} and Y_{t-1} are conditionally independent, given Y_t when independence holds for each partial table within which Y_t is fixed. That is, if

$$\pi_{ij|k} = \pi_{i+|k}\pi_{+j|k}, \text{ for all i,j and k.}$$

For joint probabilities in the table:

$$\pi_{ijk} = \pi_{i+k}\pi_{+jk}/\pi_{++k} \quad \text{for all i,j and k.}$$

The conditional independence of Y_{t-2} and Y_{t-1} are conditionally independent, given Y_j, is the loglinear model:

$$\log \mu_{ijk} = \lambda + \lambda_i^{Y_{t-2}} + \lambda_j^{Y_{t-1}} + \lambda_k^{Y_t} + \lambda_{ik}^{Y_{t-2}Y_t} + \lambda_{jk}^{Y_{t-1}Y_t}.$$

This is a weaker condition than mutual or joint independence. Mutual independence implies that Y_t is jointly independent of Y_{t-2} and Y_{t-1}, which implies that Y_{t-2} and Y_{t-1} are conditionally independent.

The general loglinear model for a three way table is

$$\log \mu_{ijk} = \lambda + \lambda_i^{Y_{t-2}} + \lambda_j^{Y_{t-1}} + \lambda_k^{Y_t} + \lambda_{ik}^{Y_{t-2}Y_t} + \lambda_{ij}^{Y_{t-2}Y_{t-1}} + \lambda_{jk}^{Y_{t-1}Y_t} + \lambda_{ijk}^{Y_{t-2}Y_{t-1}Y_t}$$

The number of cell counts is equal to the number of parameters in the above model. Hence this is called a saturated model.

11.6 Likelihood Functions and Likelihood Estimates

The joint Poisson probability for two way table for all counts $\{Y_{ij} = n_{ij}\}$ is

$$L = \prod_i \prod_j \frac{e^{-\mu_{ij}} \mu_{ij}^{n_{ij}}}{n_{ij}!}.$$

Taking log and differentiating with respect to the parameters, we can show

$$\frac{\partial \ln L(\mu)}{\partial \lambda} = n - \mu_{..} = 0, \quad \frac{\partial \ln L(\mu)}{\partial \lambda_i^{Y_{j-1}}} = n_{i.} - \mu_{i.} = 0, \quad \frac{\partial \ln L(\mu)}{\partial \lambda_j^{Y_j}} = n_{.j} - \mu_{.j} = 0.$$

Similarly, the joint Poisson probability for three way table for all counts $\{Y_{ijk} = n_{ijk}\}$ is

$$L = \prod_i \prod_j \prod_k \frac{e^{-\mu_{ijk}} \mu_{ij}^{n_{ijk}}}{n_{ijk}!}.$$

Taking log and differentiating with respect to the parameters, we can show

$$\frac{\partial \ln L(\mu)}{\partial \lambda} = n - \mu_{...} = 0, \quad \frac{\partial \ln L(\mu)}{\partial \lambda_i^{Y_{j-2}}} = n_{i..} - \mu_{i..} = 0, \quad \frac{\partial \ln L(\mu)}{\partial \lambda_j^{Y_{j-1}}} = n_{.j.} - \mu_{..j.} = 0.$$

We can also obtain the equations for other parameters.

Generalization of Test Based on log Linear Models

We know that a linear link function takes the form

$$\mu_i = \sum_{j=1}^{p} x_{ij}\beta_j,$$

where the β_j's are parameters whose values are usually unknown and have to be estimated from data. In other words,

$$E(Y_i) = \mu_i = \sum_{j=1}^{p} x_{ij}\beta_j \quad ; i=1,2,\ldots,n,$$

where x_{ij} is the value of the jth covariate for observation i. We have shown that we may write

$$\mu = X\beta$$

where μ is nx1 vector of means, X is nxp matrix of values of covariates and β is px1 vector of parameters. Alternatively, we can employ log or logit link functions for analyzing the Markov models. In a log linear form the log-link is $\eta = \log\mu$, with its inverse $\mu = e^{\eta}$. Similarly, the logit link function can be expressed as $\eta = \log\mu/(1-\mu)$.

In the previous section, we have shown for first and second order Markov models that

$$P\{Y_t = j | Y_0 = i_0, ..., Y_{t-2} = i_{t-2}, Y_{t-1} = i\} = P\{Y_t = j | Y_{t-1} = i\},$$

$$P\{Y_t = j | Y_0 = i_0, ..., Y_{t-2} = i, Y_{t-1} = j\} = P\{Y_t = k | Y_{t-2} = i, Y_{t-1} = j\}.$$

These equations can be expressed as functions of covariates as well:

$$P\{Y_t = k | Y_0 = i_0, ..., Y_{t-2} = i_{t-2}, Y_{t-1} = j, X = i\} = P\{Y_j = k | Y_{j-1} = j, X = i\},$$

$$P\{Y_t = l | Y_0 = i_0, ..., Y_{t-2} = j, Y_{t-1} = k, X = i\} = P\{Y_t = k | Y_{t-2} = j, Y_{t-1} = k, X = i\},$$

where X is a single binary covariate with X=0,1. Here we are using a single binary covariate to show the extension for both log and logit link functions. We can generalize this further for more categorical variables for log link and more categorical, discrete and continuous variables for logit link function without loss of any generality.

The model for first order Markov chain for log link function is based on a three way table:

$$\log\mu_{ijk} = \lambda + \lambda_i^X + \lambda_j^{Y_{t-1}} + \lambda_k^{Y_t}.$$

The second order Markov model with a single covariate for log link function is: based on a four way table:

$$\log\mu_{ijkl} = \lambda + \lambda_i^X + \lambda_j^{Y_{t-2}} + \lambda_k^{Y_{t-1}} + \lambda_l^{Y_t}.$$

These models are independence model which can be extended easily for models with association among the variables.

11.7 Parameter Estimation for Logit Link Function

Let us consider a Markov model of order r, then

$$P\{y_t = 1 | y_{t-1}, ..., y_{t-r}, x\} = e^{\lambda' z} / (1 + e^{\lambda' z})$$

where

$$z = (1, y_{t-r}, ..., y_{t-1}, x)' = (z_0, z_1, ..., z_{r+1})' \text{ and}$$

$$\lambda = (\beta_0, \beta_1, ..., \beta_r, \gamma)' = (\lambda_0, \lambda_1, ..., \lambda_{r+1})'.$$

In a generalized linear model, the log likelihood function is

$$l(y_t, \lambda) = \sum_{i=1}^{n} [\{z_{ij}\theta_i - b(\theta_i)\} / a(\phi) + c(z_{ij}, \phi)].$$

In case of Bernoulli distribution, the parameters are

$\theta = \ln\left[\pi/(1-\pi)\right]$, $\pi = e^{\theta}/(1+e^{\theta})$, $b(\theta) = -\ln(1-\pi)$, $a(\phi)=1$, $c(z,\phi)=0$, and

$$E(y) = \left(\frac{1}{1-\pi}\right)\pi(1-\pi) = \pi.$$

For the canonical link, we have $\eta_i = \theta_i = g[E(y_i)] = g(\mu_i) = \lambda' z_i$, ($\mu_i = \pi_i$) and therefore,

$$\frac{\partial l}{\partial \lambda} = \frac{\partial l}{\partial \theta_i}.\frac{\partial \theta_i}{\partial \lambda} = \frac{1}{a(\phi)}\sum_{i=1}^{n}[z_{ij} - \frac{db(\theta_i)}{d\theta_i}]z_{ij} = \frac{1}{a(\phi)}\sum_{i=1}^{n}[z_{ij} - \mu_{ij}]z_{ij}.$$

Here, $a(\phi)$ is a constant, so this equation becomes

$$\sum_{i=1}^{n}[z_{ij} - \mu_{ij}]z_{ij} = 0.$$

The null hypothesis $H_0 : \lambda_j = 0$, j=1,2,….,r , in the model

$$P\left\{y_t = 1 \middle| y_{t-1},\ldots,y_{t-r},x\right\} = e^{\lambda' z}/(1+e^{\lambda' z})$$

indicates whether r-th order model should be fitted or not. We can start with first order model first, if it shows significant association, we can continue for second order, and so on.

11.8 APPLICATIONS

We have considered the same HRS data on mobility of elderly population for the period 1992-2004 to display the fitting of models in this chapter. We have defined the outcome variable based on the difficulty in mobility of elderly population. For the application of the models, we have considered 0= no difficulty, 1= difficulty in one or more of the five tasks.

Table 11.1 shows the transition counts for the first order Markov model. Based on these transition counts, we have fitted the log linear model for the first order which is displayed in Table 11.2. For the first order model, we observe that both the consecutive outcomes are positively associated with the log-mean under the independence model. However, the model with interaction term shows significant positive association but the main effects for two consecutive outcomes become negative under the saturated model. Table 11.3 displays the transition counts for a second order Markov model. The second order independence model, based on three consecutive outcomes, displays that all the main-effects are negatively associated. The log linear model for second order shows that all the interaction terms are positive while the main effects are positive. In the second order model involving three consecutive outcomes, after including the interaction terms Y_1Y_2, Y_1Y_3, and Y_2Y_3 we observe that the main effects produce negative association but the interaction terms show positive association with the log-mean. The measure of deviance is also very small indicating a good fit. The results do not vary much if we consider the saturated model.

Table 11.1. Pooled Transition Counts in Mobility Index among Elderly during 1992-2004

	Transition Count		
State	0	1	Total
0	22461	5621	28082
1	3733	12636	16369

Table 11.2. Estimates of Parameters of Loglinear Model Assuming Poisson Distribution and Log Link Function for Two Way Table

Variables	Estimate	S.E.	Chi-square	p-value
Model I				
Intercept	8.8133	0.0072	1839780	0.0001
Y_1 (0)	0.5397	0.0098	3012.57	0.0001
Y_2 (0)	0.3610	0.0096	1401.91	0.0001
Deviance (Value/DF)	14502.53	(14502.53)		
Log Likelihood	371921.77			
Model III				
Intercept	9.4443	0.0089	1127067	0.0001
Y_1 (0)	-0.8100	0.0160	2552.74	0.0001
Y_2 (0)	-1.2193	0.0186	4284.43	0.0001
Y_1 (0)*Y_2 (0)	2.6046	0.0239	11913.3	0.0001
Deviance (Value/DF)				
Log Likelihood	379173.03			

Table 11.3. Pooled Transition Counts for Second Order

	Transition Count		
State	0	1	Total
0 0	15687	2833	18520
0 1	1825	2589	4414
1 0	1535	1399	2934
1 1	1263	8098	9361

The logit models are fitted for first, second and third orders taking previous outcomes as covariates. The results are summarized in Table 11.5. The second order model shows that two previous outcomes exert positive impact on the third outcome (Model II). Model III includes the interaction term Y_1Y_2 but it does not change the relationships substantially. The interaction term is not statistically significant. The Model IV includes the main effects of a third order Markov model and all the previous outcomes show positive association with the index outcome. After inclusion of interactions Y_1Y_2, Y_1Y_3, and Y_2Y_3 in the third order model (Model V), we observe that the main effects due to previous outcomes change slightly but still remain positively associated. However, the interaction terms Y_1Y_2 and Y_1Y_3 show significant negative and positive associations respectively. It is interesting to note that if we include the interaction term $Y_1 Y_2 Y_3$ in the third order model (Model VI) then all the main effects and interaction terms become significant. The sign of the main effects for three previous outcomes remain similar but the interactions $Y_1 Y_2$, $Y_1 Y_3$, and $Y_2 Y_3$ show negative associations and $Y_1 Y_2 Y_3$ displays statistically significant positive association.

Table 11.4. Estimates of Parameters of the Loglinear Model Assuming Poisson Distribution and Log Link Function for Three Way Table

Variables	Estimate	S.E.	Chi-square	p-value
Model VI				
Intercept	7.6187	0.0128	353529	0.0001
Y_1 (0)	0.3085	0.0108	818.46	0.0001
Y_2 (0)	0.4431	0.0109	1646.71	0.0001
Y_3 (0)	0.6234	0.0112	3110.86	0.0001
Deviance (Value/DF)	25881.58	(6470.39)		
Log Likelihood	263238.37			
Model V				
Intercept	9.0031	0.0109	681851	0.0001
Y_1 (0)	-1.8861	0.0264	5119.48	0.0001
Y_2 (0)	-1.7815	0.0256	4854.13	0.0001
Y_3 (0)	-1.1559	0.0210	3026.43	0.0001
Y_1 (0)*Y_2 (0)	2.0204	0.0291	4811.40	0.0001
Y_1 (0)*Y3 (0)	1.5647	0.0302	2686.85	0.0001
Y2 (0)*Y_3 (0)	1.8940	0.0300	3998.10	0.0001
Deviance (Value/DF)	3.3444	(3.3444)		
Log Likelihood	276177.48			
Model VI				
Intercept	8.9994	0.0111	655847	0.0001
Y_1 (1)	-1.8581	0.0303	3772.33	0.0001
Y_2 (1)	-1.7559	0.0290	3677.80	0.0001
Y_3 (1)	-1.1403	0.0226	2551.10	0.0001
Y_1 (1)*Y_2 (1)	1.9509	0.0478	1668.19	0.0001
Y_1 (1)*Y3 (1)	1.5084	0.0430	1230.30	0.0001
Y2 (1)*Y_3 (1)	1.8459	0.0397	2160.00	0.0001
Y_1 (1)*Y_2 (1)* Y_3 (1)	0.1103	0.0603	3.35	0.0673
Deviance (Value/DF)				
Log Likelihood	276179.16			

Table 11.5. Estimates of Parameters of Logistic Regression Model Assuming Binomial Distribution and Logit Link Function by Taking Previous Outcomes as Covariates

Variables	Estimate	S.E.	Chi-square	p-value
Model I				
Intercept	1.2195	0.0186	4285.70	0.0001
Y_1 (0)	-2.6044	0.0239	11913.1	0.0001
Deviance (Value/DF)	45702.28	(1.0281)		
Log Likelihood	-22851.14			

Table 11.5. Continued

Variables	Estimate	S.E.	Chi-square	p-value
Model II				
Intercept	-1.6930	0.0190	7954.13	0.0001
Y_1 (1)	2.0234	0.0288	4947.38	0.0001
Y_2 (1)	1.5400	0.0297	2689.12	0.0001
Deviance (Value/DF)	33977.69	(0.9505)		
Log Likelihood	-16988.85			
Model III				
Intercept	-1.7035	0.0203	7072.99	0.0001
Y_1 (1)	2.0574	0.0365	3183.31	0.0001
Y_2 (1)	1.5837	0.0413	1473.66	0.0001
Y_1 (1)*Y_2 (1)	-0.0904	0.0593	2.32	0.1275
Deviance (Value/DF)	33975.37	(0.9504)		
Log Likelihood	-16987.69			
Model VI				
Intercept	-1.7876	0.0226	6278.33	0.0001
Y_1 (1)	1.7790	0.0339	2750.73	0.0001
Y_2 (1)	1.2686	0.0358	1253.43	0.0001
Y_3 (1)	0.9260	0.0365	643.39	0.0001
Deviance (Value/DF)	25649.19	(0.9286)		
Log Likelihood	-12824.59			
Model V				
Intercept	-1.7955	0.0250	5142.93	0.0001
Y_1 (1)	1.7785	0.0468	1444.19	0.0001
Y_2 (1)	1.3778	0.0544	640.42	0.0001
Y_3 (1)	0.8889	0.0585	231.10	0.0001
Y_1 (1)*Y_2 (1)	-0.1532	0.0721	4.51	0.0337
Y_1 (1)*Y3 (1)	0.1815	0.0749	5.88	0.0153
Y2 (1)*Y_3 (1)	-0.1051	0.0749	1.97	0.1606
Deviance (Value/DF)	25639.49	(0.9284)		
Log Likelihood	-12819.75			
Model VI				
Intercept	-1.8293	0.0257	5060.59	0.0001
Y_1 (1)	1.9065	0.0498	1466.72	0.0001
Y_2 (1)	1.5584	0.0588	702.11	0.0001
Y_3 (1)	1.1009	0.0631	304.10	0.0001
Y_1 (1)*Y_2 (1)	-0.5704	0.0890	41.04	0.0001
Y_1 (1)*Y3 (1)	-0.3579	0.1004	12.70	0.0004
Y2 (1)*Y_3 (1)	-0.6731	0.1038	42.06	0.0001
Y_1 (1)*Y_2 (1)* Y_3 (1)	1.1389	0.1468	60.17	0.0001
Deviance (Value/DF)	25580.05	(0.9262)		
Log Likelihood	-12790.02			

11.9 SUMMARY

This chapter demonstrates the relationship between the first or higher order Markov models and the generalized linear models. It is shown that the Markov models can be expressed as log-linear model for examining the relationships among the outcomes at different time points. The proposed models can be employed for testing the independence of the outcomes for the first or higher order. In other words, these models provide an alternative to the tests for adequacy of Markov chains. Avery and Henderson (1999) provided the use of log-linear models to fit low order Markov chain models to discrete state series. This chapter generalizes the fitting of such models as well as the links with the generalized linear models and logistic regression models are also discussed. Agresti (2002) provided some important guidelines for analyzing categorical repeated measures. These models are easy to compute as well as to interpret. This can be used as another alternative to the approach suggested by Jones and Crowley (1992).

Chapter 12

MARGINAL AND CONDITIONAL MODELS

12.1 INTRODUCTION

We have shown the link between generalized linear model and Markov models in Chapter 11. The log linear and logit link function based relationships are highlighted in the previous chapter. This chapter is a further extension of Chapter 11. In addition, the differences between marginal and conditional models are also illustrated here, so that the users understand the strength and weakness of the model they are using. The marginal model based on the work of Azzalini (1994) is shown here and the generalized estimating equations (GEE) are also discussed. Then conditional approach introduced by Muenz and Rubinstein (1985), Bonney (1987) and Islam et al. (2004) and Islam and Chowdhury (2006) are discussed.

12.2 GENERALIZED LINEAR MODEL

The repeated measures data can be analyzed by employing both marginal and conditional models. The marginal models include Generalized Estimating Equations (GEE) and models based on marginal distribution of the binary responses in serially correlated binary data. The GEE approach has been developed for both population averaged or subject specific models and the marginal model proposed by Azzalini (1994) relates Markov conditional probabilities in terms of marginal rather than conditional sense, relating covariates to the mean value of the process, independently of the association parameters. In addition, Diggle et al (2002) showed a method of employing marginal model with lagged response variables. Lindsey and Lambert (1998) observed that it is erroneous to include lagged response variables in a marginal model. They also argued that marginal coefficients have the same interpretation as coefficients from a cross sectional analysis is true, but in that case the marginal models fail to utilize the relationships in explaining the repeated responses in terms of changes over time. In other words, the limitations of cross sectional analysis are not overcome by using marginal models for analyzing repeated measures data. On the other hand, the conditional models are sensitive to the changes in the response variables over time and thus reveal the changing pattern in repeated responses in a more realistic way and the difference between cross sectional and longitudinal measures become more evident.

Likelihood Functions for Generalized Linear Models

We assume that each component of Y has a distribution in the exponential family, taking the form

$$f_Y(y;\theta,\phi) = e^{\{(y\theta - b(\theta))/a(\phi) + c(y,\phi)\}} \tag{12.1}$$

for some specific functions a(.), b(.) and c(.). If ϕ is known, this is an exponential family model with canonical parameter θ. It may or may not be a two-parameter exponential family with canonical parameter θ. Also, it may or may not be a two parameter exponential family if ϕ is unknown.

For any binomial variable y, the probability mass function can be expressed as

$$f_Y(y;\theta,\phi) = \binom{n}{y}\pi^y(1-\pi)^{n-y} = e^{\{\ln\binom{n}{y} + y\ln\pi + (n-y)\ln(1-\pi)\}}$$

$$= e^{\{\ln\binom{n}{y} + y\ \ln\pi + n\ln(1-\pi) - y\ \ln(1-\pi)\}} = e^{\{y\ \ln\left[\frac{\pi}{1-\pi}\right] + n\ \ln(1-\pi) + \ln\binom{n}{y}\}},$$

which is of the generalized linear model form given by (12.1). Therefore, for the binomial distribution

$$\theta = \ln[\pi/(1-\pi)],\ \pi = e^{\theta}/(1+e^{\theta}),\ b(\theta) = -n\ln(1-\pi),\ a(\phi) = 1,$$

$$c(y,\phi) = \ln\binom{n}{y},\ E(y) = \frac{db(\theta)}{d\theta} = \frac{db(\theta)}{d\pi}.\frac{d\pi}{d\theta},\ \text{where}$$

$$\frac{d\pi}{d\theta} = \frac{e^{\theta}}{1+e^{\theta}} - \left[\frac{e^{\theta}}{1+e\theta}\right]^2 = \pi(1-\pi).$$

Therefore,

$$E(y) = [n/(1-\pi)]\pi(1-\pi) = n\pi. \tag{12.2}$$

We recognize (12.2) as the mean of the binomial distribution. Also,

$$Var(y) = \frac{dE(y)}{d\theta} = \frac{dE(y)}{d\pi}\frac{d\pi}{d\theta} = n\pi(1-\pi), \tag{12.3}$$

(12.3) being obtained from (12.2).

12.3 Binomial Logit Model for Binary Data: Conditional Approach

If the response variable is binary, then we can express the variable as Y=1, for success, and Y=0 for failure. Then for Bernoulli distribution, we can express

$$P(Y=1) = \pi, P(Y=0) = 1-\pi \text{ for which } E(Y) = \mu = \pi.$$

This is a special case of binomial distribution with n=1.

Then we can see that

$$\theta = \log[\pi/(1-\pi)] = \log[\mu/(1-\mu)].$$

The natural parameter θ is the log odds of response 1 which is called logit of π. The model using this link is

$$\eta_i = g(\mu_i) = \log[\mu_i/(1-\mu_i)] = \sum_{j=1}^{p} \beta_j x_{ij}, \text{ i=1,2,...,n.}$$

This is the canonical link and the model is called the logit model.

12.4 Binomial GLM for 2x2 Contingency Table

Let us consider a 2x2 table for variables X and Y. If Y is considered as a dependent variable with Y=1 for occurrence of event, and Y=0 otherwise, and X=1 for exposure and X=0 for non-exposure, then the canonical link is

$$\begin{aligned} P\left[Y_i=1 \middle| X=x_i\right] &= \pi(x_i) = \frac{e^{\beta_0+\beta_1 x_i}}{1+e^{\beta_0+\beta_1 x_i}}, \\ P\left[Y_i=0 \middle| X=x_i\right] &= 1-\pi(x_i) = 1-\frac{e^{\beta_0+\beta_1 x_i}}{1+e^{\beta_0+\beta_1 x_i}} = \frac{1}{1+e^{\beta_0+\beta_1 x_i}}. \end{aligned} \tag{12.4}$$

Now we obtain the following odds for X=1 or X=0:

If X=1, then from (12.4) we get

$$\begin{aligned} P\left[Y_i=1 \middle| X=1\right] &= \pi(1) = \frac{e^{\beta_0+\beta_1}}{1+e^{\beta_0+\beta_1}}, \\ P\left[Y_i=0 \middle| X=1\right] &= 1-\pi(1) = 1-\frac{e^{\beta_0+\beta_1}}{1+e^{\beta_0+\beta_1}} = \frac{1}{1+e^{\beta_0+\beta_1}}, \\ \frac{\pi(1)}{1-\pi(1)} &= \frac{\dfrac{e^{\beta_0+\beta_1}}{1+e^{\beta_0+\beta_1}}}{\dfrac{1}{1+e^{\beta_0+\beta_1}}} = e^{\beta_0+\beta_1}, \\ \log it\left[\pi(1)\right] &= \log\frac{\pi(x_i=1)}{1-\pi(x_i=1)} = \beta_0+\beta_1. \end{aligned} \tag{12.5}$$

Similarly, for X=0 we get

$$P\left[Y_i = 1 \middle| X = 0\right] = \pi(0) = \frac{e^{\beta_0}}{1+e^{\beta_0}},$$

$$P\left[Y_i = 0 \middle| X = 0\right] = 1 - \pi(0) = 1 - \frac{e^{\beta_0}}{1+e^{\beta_0}} = \frac{1}{1+e^{\beta_0}},$$

$$\frac{\pi(0)}{1-\pi(0)} = \frac{\frac{e^{\beta_0}}{1+e^{\beta_0}}}{\frac{1}{1+e^{\beta_0}}} = e^{\beta_0}, \tag{12.6}$$

$$\log it\left[\pi(0)\right] = \log\frac{\pi(0)}{1-\pi(0)} = \beta_0.$$

Hence, from (12.5) and (12.6),

$$\log it\left[\pi(1)\right] - \log it\left[\pi(0)\right] = \beta_1. \tag{12.7}$$

As (12.7) shows, the difference between two log odds gives the estimate of β_1.

This can be generalized for a vector of p risk factors, X, as follows:

$$P\left[Y_i = 1 \middle| X = x\right] = \pi(x) = \frac{e^{x'\beta}}{1+e^{x'\beta}},$$

$$P\left[Y_i = 0 \middle| X = x\right] = 1 - \pi(x) = 1 - \frac{e^{x'\beta}}{1+e^{x'\beta}} = \frac{1}{1+e^{x'\beta}}.$$

where $x' = (1, x_1, ..., x_p)$ and $\beta = (\beta_0, \beta_1, ..., \beta_p)$ and the logit function is

$$\log it\left[\pi(x)\right] = x'\beta.$$

The full information maximum likelihood for the Bernoulli distribution is

$$L(\mu | y_1, ..., y_n) = \prod_{i=1}^{n} \exp\{y_i \ln[\mu_i/(1-\mu_i)] + \ln(1-\mu_i)\},$$

and $E(y_i) = \pi_i = \mu_i$, and $Var(y_i) = \pi_i(1-\pi_i) = \mu_i(1-\mu_i)$.

Then the link function is

$$g(\mu_i) = \log it\left[\mu_i\right] = \ln\frac{\mu_i}{1-\mu_i} = x_i'\beta, \text{ where } x_i' = (1, x_{i1}, ..., x_{ip}).$$

A limited information maximum likelihood can be expressed for the following probability mass function:

$$f_Y(y;\theta,\phi) = e^{\{(y\theta - b(\theta))/a(\phi) + c(y,\phi)\}}$$

where $E(y) = b'(\theta) = \mu$, $Var(y) = b''(\theta)a(\phi)$.

The log-likelihood of the above exponential family is

$$\ln L\left(\theta,\phi\middle|y_1,y_2,...,y_n\right)=\sum_{i=1}^{n}\left\{\frac{y_i\theta_i-b(\theta_i)}{a(\phi)}+c(y_i,\phi)\right\}.$$

Using the chain rule we can derive the estimating equation:

$$\frac{\partial \ln L}{\partial \beta}=\left[\left(\frac{\partial \ln L}{\partial \theta}\right)\left(\frac{\partial \theta}{\partial \mu}\right)\left(\frac{\partial \mu}{\partial \eta}\right)\left(\frac{\partial \eta}{\partial \beta_j}\right)\right]_{p\times 1}=\left[\sum_{i=1}^{n}\left(\frac{y_i-b'(\theta_i)}{a(\phi)}\right)\left(\frac{1}{V(\mu_i)}\right)\left(\frac{\partial \mu}{\partial \eta}\right)_i(x_{ji})\right]_{p\times 1}$$ Ac

$$=\left[\sum_{i=1}^{n}\frac{y_i-\mu_i}{a(\phi)V(\mu_i)}\left(\frac{\partial \mu}{\partial \eta}\right)_i(x_{ji})\right]_{p\times 1}=(0)_{p\times 1}.$$

cording to Wedderburn (1974), we can view the mean and variance functions as part of the estimating equations under limited information maximum quasi likelihood approach. A sandwich estimate of the following form can be employed:

$$V_S(\hat{\beta})=\hat{V}_H^{-1}(\hat{\beta})\hat{B}(\hat{\beta})\hat{V}_H^{-1}(\hat{\beta})$$

where

$$\hat{V}_H(\hat{\beta})=\left[-\frac{\partial^2 \ln L}{\partial \beta_u \partial \beta_v}\right]_{pxp}^{-1},\ \hat{B}(\hat{\beta})=\left[\sum_{i=1}^{n}x_i'\left\{\frac{y_i-\mu_i}{V(\hat{\mu}_i)}\left(\frac{\partial \mu}{\partial \eta}\right)_i\hat{\phi}\right\}^2 x_i\right]_{p\times p}.$$

12.5 Marginal Logistic Regression

Now let us consider the relationship between Y and X and a third variable, Z. For the time being, we consider all the variables binary. Using P(Y=1/X) for Z=0 and Z=1, we obtain the following models:

$$P\left[Y_i=1\middle|X,Z\right]=\pi(x_i,z)=\frac{e^{\beta_0(z)+\beta_1(z)x_i}}{1+e^{\beta_0(z)+\beta_1(z)x_i}},$$

$$P\left[Y_i=0\middle|X,Z\right]=1-\pi(x_i,z)=1-\frac{e^{\beta_0(z)+\beta_1(z)x_i}}{1+e^{\beta_0(z)+\beta_1(z)x_i}}=\frac{1}{1+e^{\beta_0(z)+\beta_1(z)x_i}}. \tag{12.8}$$

For Z=0 and for Z=1 the models are

$$P\left[Y_i=1\middle|X,Z=0\right]=\pi(x_i,0)=\frac{e^{\beta_0(0)+\beta_1(0)x_i}}{1+e^{\beta_0(0)+\beta_1(0)x_i}}, \tag{12.9}$$

$$P\left[Y_i=1\middle|X,Z=1\right]=\pi(x_i,1)=\frac{e^{\beta_0(1)+\beta_1(1)x_i}}{1+e^{\beta_0(1)+\beta_1(1)x_i}}.$$

Using (12.8) and (12.9) we obtain the logit expressions as follows:

$$\frac{\pi(x_i, Z)}{1-\pi(x_i, Z)} = \left[\frac{e^{\beta_0(z)+\beta_1(z)x_i}}{1+e^{\beta_0+\beta_1(z)x_i}} \Bigg/ \frac{1}{1+e^{\beta_0(z)+\beta_1(z)x_i}} \right] = e^{\beta_0(z)+\beta_1(z)x_i}.$$

Hence,

$$\log it\left[\pi(x_i, Z)\right] = \log\frac{\pi(x_i, Z)}{1-\pi(x_i, Z)} = \beta_0(z) + \beta_1(z)x_i. \qquad (12.10)$$

In other words, we have two logit functions, namely,

$$\log it\left[\pi(x_i, Z=0)\right] = \log\frac{\pi(x_i, Z=0)}{1-\pi(x_i, Z=0)} = \beta_0(0) + \beta_1(0)x_i,$$

$$\log it\left[\pi(x_i, Z=1)\right] = \log\frac{\pi(x_i, Z=1)}{1-\pi(x_i, Z=1)} = \beta_0(1) + \beta_1(1)x_i.$$

Guo and Geng (1995) showed that the logistic regression coefficients $\beta_1(z)$ are collapsible over Z if the marginal logistic regression

$$\log it\left[\pi(x_i)\right] = \log[\pi(x_i)/(1-\pi(x_i))] = \beta_0 + \beta_1 x_i$$

can be obtained after collapsing the background Z and if $\beta_1(z) = \beta_1$ for all z.

We can generalize the definition (12.10) for any dichotomous Y, letting X be a continuous, discrete or mixed random vector of p risk factors (first variable in X is 1 corresponding to β_0) and Z be a discrete random variable with levels z=1,2,…,I. Then the logit function can be defined as

$$\log it\left[\pi(x, z)\right] = \log[\pi(x, z)/\{1-\pi(x, z)\}] = x'\beta(z).$$

The marginal logistic regression is then

$$\log it\left[\pi(x)\right] = \log[\pi(x)/\{1-\pi(x)\}] = x'\beta$$

assuming that $\beta(z) = \beta$. The collapsibility implies that the conditional and marginal coefficients are the same, and the marginal logit is still linear. It is worth noting that if the coefficients $\beta(z)$ are simply collapsible, then the background variable, Z, is called a non-confounding covariate. If these are simply collapsible, then we can investigate the marginal model by pooling the levels of Z, or even by not observing Z at all. Robinson and Jewell (1991) observed that the precision of estimates of the coefficients $\beta(z)$ may be improved by pooling the background variables if the background variable, Z, is a non-confounding covariate.

Carey, Zeger and Diggle (1993) proposed a marginal model on the basis of marginal expectation of each binary variable as well as association between pairs of outcomes in terms of explanatory variables. Let us consider z=1,2,…,Z as Z clusters and let $Y_i = (Y_{i1}, ..., Y_{in_i})'$ be an $n_i \times 1$ response vector with mean $E(Y_i) = \mu_i$ and let ψ_{ijk} be the odds ratio between responses Y_{ij} and Y_{ik} defined by

$$\psi_{ijk} = \frac{p(Y_{ij} = 1, Y_{ik} = 1)p(Y_{ij} = 0, Y_{ik} = 0)}{p(Y_{ij} = 1, Y_{ik} = 0)p(Y_{ij} = 0, Y_{ik} = 1)}.$$

Then the marginal model can be specified as

$$\log it\left[\mu_{ij}\right] = \log[\mu_{ij} /(1 - \mu_{ij})] = x'_{ij}\beta_j$$

where x_{ij} is a px1 vector of explanatory variables associated with Y_{ij} and β_j s are vector of regression coefficients to be estimated. Similarly,

$$\log\left[\psi_{ijk}\right] = z'_{ijk}\alpha$$

where z_{ijk} is a qx1 vector of covariates which specifies the association between Y's and α is a qx1 vector of association parameters to be estimated. This requires generalized estimating equation approach for estimating the parameters.

12.6 GENERALIZED ESTIMATING EQUATIONS

We have introduced the generalized linear models in the beginning of this chapter for binary outcomes. It is interesting to note that the estimating equations for GLMs has its root in log-likelihood based upon the exponential family of distributions. Following the work of Wedderburn (1974), the utility of this estimating equations can be extended further outside the implications of log-likelihood. The estimating equation methods are related to the quasilikelihood methods in that there are no parametric assumptions.

The term generalized estimating equations (GEE) indicates that an estimating equation, not based on log-likelihood, is a generalization of another estimating equation obtained by second order variance components directly into a pooled estimating equation.

There are two types of models: (i) population averaged (PA) and (ii) subject specific (SS). Let us consider that for a given outcome Y_{ij}, we have a px1 vector of covariates x_{ij}, and the associated parameter vector is β. We can also define a qx1 vector of covariates z_{ij} associated with the random effect ν_i. The conditional expectation for subject specific model is given by

$$\mu_{ij}^{SS} = E\left(Y_{ij} \left| \nu_i\right.\right).$$

The responses for any panel i can be expressed as follows for subject specific:

$$g\left(\mu_{ij}^{SS}\right) = x_{ij}\beta^{SS} + x_{ij}\nu_i,\ V\left(Y_{ij} \left| \nu_i\right.\right) = V\left(\mu_{ij}^{SS}\right).$$

We can focus on the distribution of the random effects as the source of nonindependence, or we can consider the marginal expectation of the outcome

$$\mu_{ij}^{PA} = E[E\left(Y_{ij} \left| \nu_i\right.\right)]$$

so that the responses are characterized by

$$g\left(\mu_{ij}^{PA}\right)=x_{ij}\beta^{PA},\ V\left(Y_{ij}\left|v_i\right.\right)=V\left(\mu_{ij}^{PA}\right)a(\phi).$$

Thus the marginal expectation is the average response for observations sharing the same covariates across all panels.

The logit functions and the odds ratios for subject specific and population averaged are:

$$\log it\left[\pi(x_{ij},v_i)\right]=\beta_0^{SS}+\beta_1^{SS}x_{ij}+v_i \text{ and}$$

$$\log it\left[\pi(x_{ij})\right]=\beta_0^{PA}+\beta_1^{PA}x_{ij}.$$

Similarly, the odds ratios are

$$OR^{SS}=\frac{P(Y_{ij}=1\left|X_{ij}=1,v_i\right.)/P(Y_{ij}=0\left|X_{ij}=1,v_i\right.)}{P(Y_{ij}=1\left|X_{ij}=0,v_i\right.)/P(Y_{ij}=0\left|X_{ij}=0,v_i\right.)}=e^{\beta_1^{SS}},$$

$$OR^{PA}=\frac{P(Y_{ij}=1\left|X_{ij}=1\right.)/P(Y_{ij}=0\left|X_{ij}=1\right.)}{P(Y_{ij}=1\left|X_{ij}=0\right.)/P(Y_{ij}=0\left|X_{ij}=0\right.)}=e^{\beta_1^{PA}}.$$

The Population Averaged Generalized Estimating Equations (PA-GEE)

We can show the PA-GEE as a straightforward generalization of the estimating equations for GLMs

$$\psi(\beta)=\left[\left\{\sum_{i=1}^{n}\sum_{j=1}^{n_i}\frac{y_{ij}-\mu_{ij}}{a(\phi)V\left(\mu_{ij}\right)}\left(\frac{\partial\mu}{\partial\eta}\right)_{ij}x_{ijk}\right\}k=1,2,\ldots,p\right]_{p\times1}=[0]_{p\times1}$$

which can be expressed in matrix notations as

$$\psi(\beta)=\left[\left\{\sum_{i=1}^{n}x'_{ki}D\left(\frac{\partial\mu}{\partial\eta}\right)[V(\mu_i)]^{-1}\left(\frac{y_i-\mu_i}{a(\phi)}\right)\right\}k=1,2,\ldots,p\right]_{p\times1}=[0]_{p\times1}$$

where D(.) denotes a diagonal matrix. We can also show that the $V\left(\mu_i\right)$ is a diagonal matrix which can be decomposed as

$$V(\mu_i)=\left[D\left(V\left(\mu_{ij}\right)^{1/2}\right)ID\left(V\left(\mu_{ij}\right)^{1/2}\right)\right]_{n_i\times n_i}.$$

This is an independent model because estimating equation considers each observation in the panel as independent. Instead of identity matrix, if we consider a general correlation matrix, then

$$V(\mu_i)=\left[D\left(V\left(\mu_{ij}\right)^{1/2}\right)R(\alpha)D\left(V\left(\mu_{ij}\right)^{1/2}\right)\right]_{n_i\times n_i}$$

where $R(\alpha)$ is the correlation matrix which is a function of parameter vector α. The following correlation structures can be considered:

(i) Exchangeable Correlation: In this case, the elements of the correlation matrix are: $R_{uv} = 1$, if u=v, and $R_{uv} = \alpha$, otherwise.

(ii) Autoregressive Correlation: The correlation structure between Y_{ij} and $Y_{ij'}$ is represented by $\alpha^{|j-j'|}$.

(iii) Stationary Correlation: The correlation structure is represented by $R_{uv} = \alpha_{|u-v|}$, if $|u-v| \le k$, $R_{|u-v|} = 0$, otherwise.

(iv) Unstructured Correlation: The working correlation matrix for unstructured correlation is $R = \alpha$.

The Subject Specific Generalized Estimating Equations (SS-GEE)

For a SS-GEE, we need to find: (i) a distribution for the random component, (ii) expected value which depends on the link function and the distribution of the random component, and (iii) a variance that involves the usual variance and the random effect.

To estimate the SS-GEE, we may consider the following equations:

$$\psi(\beta,\alpha) = \left[\left\{\sum_{i=1}^{n} x'_{ki} D\left(\frac{\partial \mu_i}{\partial \eta}\right)\left[V(\mu_i)\right]^{-1}\left(\frac{y_i - \mu_i}{a(\phi)}\right)\right\}\right] = [0]_{p\times 1},$$

$$\mu_i = \int f(v_i) g^{-1}(x_{ij}\beta^{SS} + v_i) dv_i,$$

$$\tilde{V}(\mu_i) = D\left(\frac{\partial \mu^{SS}}{\partial \eta^{SS}}\right) v_i \Sigma_v(\alpha) v'_i D\left(\frac{\partial \mu^{SS}}{\partial \eta^{SS}}\right) + \phi V\left(\mu^{SS}\right),$$

$\Sigma_v(\alpha) =$ variance matrix.

It is noteworthy that PA-GEE is used more often than SS-GEE. This is attributable to the fact that the alternative likelihood procedures are available. If the focus is on the variance structure, then SS-GEE should be considered instead of marginal model based PA-GEE.

12.7 A Model Based on Binary Markov Chain

Azzalini (1994) proposed a marginal model based on binary Markov Chain for a single stationary process $(Y_1,\ldots,Y_T)$ where Y's take values of 0 or 1 at subsequent times denoted by j=1,2,…,J.The Markov chain for any two subsequent times is represented by

$$\pi = \begin{pmatrix} \pi_{00} & \pi_{01} \\ \pi_{10} & \pi_{11} \end{pmatrix}.$$

Here $\pi_{ijt} = P\left(Y_t = j \middle| Y_{t-1} = i\right)$, i,j= 0,1. Let us denote $\mu = E\left(Y_t\right)$ for stationary and $\mu_t = E\left(Y_t\right)$ for non-stationary case. The odds ratio between successive observations is-

$$\psi = \frac{\pi_{11} / \pi_{10}}{\pi_{01} / \pi_{00}}.$$

Then the mean, $\mu = E\left(Y_t\right)$ or $\mu_t = E\left(Y_t\right)$, can be expressed as

$\mu = \mu\pi_{11} + (1-\mu)\pi_{01}$ for stationary, and

$\mu_t = \mu_{t-1}\pi_{11} + (1-\mu_{t-1})\pi_{01}$, for non-stationary case.

Azzalini (1994) showed that the transition probabilities can be expressed in terms of the odds ratios and means for successive observations, as follows:

$$\pi_{ij} = \begin{cases} \mu_t, & \text{for } \psi=1, \\ \dfrac{\delta-1+(\psi-1)(\mu_t - \mu_{t-1})}{2(\psi-1)(1-\mu_{t-1})} + \dfrac{1-\delta+(\psi-1)(\mu_t + \mu_{t-1} - 2\mu_t\mu_{t-1})}{2(\psi-1)\mu_t(1-\mu_{t-1})} & \text{for } \psi \neq 1 \end{cases}$$

where

$$\delta^2 = 1+(\psi-1)\left\{(\mu_t - \mu_{t-1})^2\psi - (\mu_t + \mu_{t-1})^2 + 2(\mu_t + \mu_{t-1}\right\}.$$

The likelihood function for a sequence of observed data $\left(y_{ij1},\ldots,y_{ijT}\right)$, i,j=0,1 is:

$$L = \prod_{t=1}^{T}\prod_{i=0}^{1}\prod_{j=0}^{1}(\pi_{ijt})^{y_{ijt}}$$

$$= \prod_{t=1}^{T}(\pi_{00t})^{(1-y_{01t})}(\pi_{01t})^{y_{01t}}(\pi_{10t})^{(1-y_{11t})}(\pi_{11t})^{y_{11t}}.$$

$$\ln L = \sum_{t=1}^{T}\left[(1-y_{01t})\ln\pi_{00t} + y_{01t}\ln\pi_{01t} + (1-y_{11t})\ln\pi_{10t} + y_{11t}\ln\pi_{11t}\right]$$

$$= \sum_{t=1}^{T}\left[y_{01t}\ln\frac{\pi_{01t}}{\pi_{00t}} + \ln\pi_{00t}\right] + \sum_{t=1}^{T}\left[y_{11t}\ln\frac{\pi_{11t}}{\pi_{10t}} + \ln\pi_{10t}\right]$$

$$= \sum_{t=1}^{T}\left[y_{01t}\operatorname{logit}\pi_{01t} + \ln\pi_{00t}\right] + \sum_{t=1}^{T}\left[y_{11t}\operatorname{logit}\pi_{11t} + \ln\pi_{10t}\right]$$

$$= \sum_{t=1}^{T}\left[y_{01t}\operatorname{logit}\mu_{01t} + \ln\mu_{00t}\right] + \sum_{t=1}^{T}\left[y_{11t}\operatorname{logit}\mu_{11t} + \ln\mu_{10t}\right].$$

Hence, the logit function for conditional means are:

$$\operatorname{logit}\mu_{01t} = x_{01t}\beta_{01},$$

$$\operatorname{logit}\mu_{11t} = x_{11t}\beta_{11}.$$

The relationship between expected value and the probabilities for marginal mean can be shown as:

$$\mu = \mu\pi_{11} + (1-\mu)\pi_{01}$$

and more specifically, for generalization, we can write

$$\mu_t = \mu_{t-1}\pi_{11} + (1-\mu_{t-1})\pi_{01}.$$

Hence, the logit function is

$$\log it\mu_t = x_t\beta.$$

The likelihood function for the marginal model, as proposed by Azzalini (1994) is

$$L = \prod_{t=1}^{T}\prod_{i=0}^{1}(\pi_{it})^{y_{it}} = \prod_{t=1}^{T}(\pi_{0t})^{(1-y_t)}(\pi_{1t})^{y_t}$$

$$\ln L = \sum_{t=1}^{T}\left[(1-y_t)\ln\pi_{0t} + y_t\ln\pi_{1t}\right] = \sum_{t=1}^{T}\left[y_t\ln\frac{\pi_{1t}}{\pi_{0t}} + \ln\pi_{0t}\right]$$

$$= \sum_{t=1}^{T}\left[y_t\log it\pi_{1t} + \ln\pi_{0t}\right] = \sum_{t=1}^{T}\left[y_t\log it\mu_t + \ln\pi_{0t}\right].$$

The estimates can be obtained from the following equations assuming equality of parameters for two logit functions:

$$\frac{\partial\ln L}{\partial\beta} = \frac{\partial\ln L}{\partial\pi_{01t}}\frac{\partial\pi_{01t}}{\partial\mu_{01t}}\frac{\partial\mu_{01t}}{\partial\beta} + \frac{\partial\ln L}{\partial\pi_{t11}}\frac{\partial\pi_{11t}}{\partial\mu_{11t}}\frac{\partial\mu_{11t}}{\partial\beta} = 0.$$

Similarly, we can express the log likelihood function for conditional model in terms of the odds ratios:

$$\ln L = \sum_{t=1}^{T}\left[(1-y_t)\ln\pi_{00t} + y_t\ln\pi_{01t} + (1-y_t)\ln\pi_{10t} + y_t\ln\pi_{11t}\right]$$

$$= \sum_{t=1}^{T}\left[y_t\ln\frac{\pi_{01t}}{\pi_{00t}} + \ln\pi_{00t}\right] + \sum_{t=1}^{T}\left[y_t\ln\frac{\pi_{11t}}{\pi_{10t}} + \ln\pi_{10t}\right]$$

$$= \sum_{t=1}^{T}\left[y_t\ln\left(\frac{\pi_{11t}/\pi_{10t}}{\pi_{01t}/\pi_{00t}}\right) + 2y_t\log it\pi_{01t} + \ln\pi_{00t} + \ln\pi_{10t}\right]$$

$$= \sum_{t=1}^{T}\left[y_t\ln\psi + 2y_t\log it\pi_{01t} + \ln\pi_{00t} + \ln\pi_{10t}\right]$$

and the solution for $\lambda = \ln\psi$ can be obtained from

$$\frac{\partial\ln L}{\partial\lambda} = \frac{\partial\ln L}{\partial\pi_{01t}}\frac{\partial\pi_{01t}}{\partial\psi}\frac{\partial\psi}{\partial\lambda} + \frac{\partial\ln L}{\partial\pi_{11t}}\frac{\partial\pi_{11t}}{\partial\psi}\frac{\partial\psi}{\partial\lambda} = 0.$$

If we consider p_1 and p_2 as the marginal proportions at time point t-1, then the probability for a Bernoulli distribution is

$\pi = \sum_{i=1}^{2} p_i\pi_i$ and the variance is $n\sum_{i=1}^{2} p_i\pi_i(1-\pi_i)$.

Hence, the marginal distribution is not Bernoulli, as employed by Azzalini (1994). Let us consider two dependent binary variables, $Y_{t-1} = Y_1$ and $Y_t = Y_2$. In addition, let us consider the association parameter, ψ. Then the joint distribution of $Y_{t-1} = Y_1$ and $Y_t = Y_2$ as shown by Lindsey and Lambert (1998) is

$$f(y_1, y_2; \nu_1, \nu_2, \psi) = \frac{1}{1+\nu_1\nu_2(\psi-1)} \nu_1^{y_1}(1-\nu_1)^{1-y_1} \nu_2^{y_2}(1-\nu_2)^{1-y_2} \psi^{y_1 y_2}, \quad \psi > 0$$

where

$$\psi = \frac{f_{11}(1-\nu_1\nu_2)}{(1-f_{11})\nu_1\nu_2}, \; f_{11} = f\,(y_1 = 1, y_2 = 1).$$

The conditional distribution of the second variable for given first variable is

$$f(y_2 | y_1 = 2-i; \nu_2, \psi) = \frac{1}{1+\nu_2(\psi^{y_1}-1)} \nu_2^{y_2}(1-\nu_2)^{1-y_2} \psi^{y_1 y_2} = \pi_i^{y_2}(1-\pi_i)^{1-y_2}, \text{ i=1,2.}$$

In the above conditional distribution,

$$\pi_1 = f(y_2 = 1 | y_1 = 1; \nu_2, \psi) = \frac{\nu_2\psi}{1+\nu_2(\psi-1)}, \; \pi_2 = f(y_2 = 1 | y_1 = 0; \nu_2, \psi) = \nu_2.$$

Lindsey and Lambert (1998) showed the marginal distribution as

$$f(y_2; \nu_1, \nu_2, \psi) = \frac{1}{1+\nu_1\nu_2(\psi-1)} \nu_2^{y_2}(1-\nu_2)^{1-y_2}[1+\nu_1(\psi^{y_2-1})].$$

The expected value and the variance are:

$$E(Y_2) = \pi = p_1\pi_1 + p_2\pi_2, \; Var(Y_2) = \pi(1-\pi).$$

It is a Bernoulli distribution but with varying probability in successive time points. Due to inclusion of π_1 and π_2 in the expected value, the link function should be

$$g(\mu) = \log\left(\frac{p_1\pi_1 + p_2\pi_2}{1-p_1\pi_1 - p_2\pi_2}\right) = x_t\beta.$$

It is clear from here that this is not the type of logit function used by Azzalini (1994). Hence, the conceptualization in the marginal models might be more complicated than the traditional logit link function. However, if we assume that $\pi_1 = \pi_2 = \pi$, then

$$g(\mu) = \log\left(\frac{p_1\pi + p_2\pi}{1-p_1\pi - p_2\pi}\right) = \log\left(\frac{\pi}{1-\pi}\right) = x_t\beta.$$

Hence, the marginal model based on the formulation of Azzalini (1994) can be employed only if $\pi_1 = \pi_2 = \pi$. In case of any Simpson's paradox problem, where conditional odds ratios differ from the marginal odds ratios, this can not provide any reliable estimate of parameters for explaining the dependence in the binary outcome data.

12.8 Models for First and Second Order Markov Models

A single stationary process $(y_{i1}, y_{i2},..., y_{ij})$ represents the past and present responses for subject i (i= 1,.2,…,n) at follow-up j (j=1,2,…, n_i). y_{ij} is the response at time t_{ij}. We can think of y_{ij} as an explicit function of past history of subject i at follow-up j denoted by $H_{ij} = \{y_{ik},\ \text{k=1,2,...,j-1}\}$. The transition models for which the conditional distribution of y_{ij}, given H_{ij}, depends on r prior observations, $y_{ij-1},..., y_{ij-r}$, is considered as the model of order r. The binary outcome is defined as y_{ij}=1, if an event occurs for the ith subject at the jth follow-up, y_{ij}=0, otherwise.

Then the first order Markov model can be expressed as

$$P(y_{ij} \middle| y_{ij-r},..., y_{ij-1}) = P(y_{ij} \middle| y_{ij-1})$$

and the corresponding transition probability matrix is given by

$$\begin{array}{cc} y_{ij-1} & y_{ij} \\ & \begin{array}{cc} 0 & 1 \end{array} \\ \begin{array}{c} 0 \\ 1 \end{array} & \begin{bmatrix} \pi_{00} & \pi_{01} \\ \pi_{10} & \pi_{11} \end{bmatrix}. \end{array}$$

Now if we consider that the process is initiated at time t_{i0} and the corresponding response is y_{i0}, then we can write the first order probabilities for the n_i follow-ups as follows:

$$P(y_{i0}, y_{i1},..., y_{in_i}) = P(y_{i0})P(y_{i1} \middle| y_{i0})P(y_{i2} | y_{i1})...P(y_{in_i} \middle| y_{in_{i-1}}).$$

We can define the conditional probabilities in terms of transition probabilities

$$\pi_{s|u} = \pi_{us} = P(y_{ij} = s \middle| y_{ij-1} = u).$$

The likelihood function can be expressed as `

$$\left\{ \prod_{u=0}^{1} \pi_u \right\} \prod_{i=1}^{n} \prod_{j=1}^{n_i} \prod_{u=0}^{1} \prod_{s=0}^{1} \pi_{us}^{y_{ij}}.$$

The maximum likelihood estimators of transition probabilities shown by Anderson and Goodman (1957) are $\hat{\pi}_{us} = n_{us} / n_{u+}$

where n_{us} = total number of transitions of type u-s, and n_{u+} =total number in state u at time t_{ij-1}.

A single stationary process $(y_{i1}, y_{i2},..., y_{ij})$ for subject i (i= 1,.2,…,n) at follow-up j (j=1,2,…, n_i) is considered. y_{ij} is the binary response at time t_{ij}, y_{ij}=0,1. It is assumed that y_{ij} is a function of past history of subject i at follow-up j denoted by $H_{ij} = \{y_{ik},\ k=j-1,j-2\}$ for second order Markov models. In other words, the transition model for order 2 presents the conditional distribution of y_{ij} given H_{ij} depending on 2 prior observations y_{ij-1}, y_{ij-2} where y_{ij-1}, y_{ij-2}=0, 1.

Then the second order Markov model can be expressed as

$$P(y_{ij} \mid H_{ij}) = P(y_{ij} \mid y_{ij-2}, y_{ij-1}) \quad \text{where } y_{ij-1}, y_{ij-2} = 0,1.$$

The transition probability matrix is

$$\begin{array}{cc|c}
y_{ij-2} & y_{ij-1} & y_{ij} \\
 & & \begin{array}{cc} 0 & 1 \end{array} \\
\begin{array}{c} 0 \\ 0 \\ 1 \\ 1 \end{array} & \begin{array}{c} 0 \\ 1 \\ 0 \\ 1 \end{array} & \begin{bmatrix} \pi_{000} & \pi_{001} \\ \pi_{010} & \pi_{011} \\ \pi_{100} & \pi_{101} \\ \pi_{110} & \pi_{111} \end{bmatrix}
\end{array}$$

12.9 Regressive Logistic Model

Another conditional model is the one proposed by Bonney (1987) and called the regressive logistic model in which both binary outcomes in previous times as well as covariates can be included. We have already discussed this model in a previous chapter. The joint mass function can be expressed as

$$P(y_{i1}, y_{i2}, \ldots, y_{in_i}; x) = P(y_{i1}; x) P(y_{i2} \mid y_{i1}; x) P(y_{i3} \mid y_{i1}; y_{i2}) \ldots P(y_{in_i} \mid y_{i1}, \ldots, y_{in_{i-1}}; x).$$

The jth logit is defined as

$$\theta_j = \ln \frac{P(y_{ij} = 1 \mid y_{i1}, y_{i2}, \ldots, y_{ij-1}; x_i)}{P(y_{ij} = 0 \mid y_{i1}, y_{i2}, \ldots, y_{ij-1}; x_i)}.$$

Bonney (1987) proposed regression model for each conditional probability as shown below

$$P(y_{ij} \mid y_{i1}, \ldots, y_{ij-1}; x_i) = \frac{e^{\theta_j y_{ij}}}{1 + e^{\theta_j}},$$

$$\theta_j = \beta_0 + \beta_1 y_{i1} + \ldots + \beta_{j-1} y_{ij-1} + \gamma_0 + \gamma_1 x_{i1} + \ldots + \gamma_p x_{ip}$$

where θ_j is the jth logit as defined above.

Now we can obtain the likelihood function as

$$L = \prod_{i=1}^{n} \prod_{j=1}^{n_i} P(y_{ij} \left| y_{i1}, \ldots, y_{ij-1}; x_i \right.) = \prod_{i=1}^{n} \prod_{j=1}^{n_i} \frac{e^{\theta_j y_{ij}}}{1+e^{\theta_j}}.$$

The estimates of the parameters can be obtained from the equations of first derivatives of log likelihood function with respect to the parameters contained in θ_j:

$$\frac{\partial \ln L}{\partial \beta} = 0, \quad \frac{\partial \ln L}{\partial \gamma} = 0.$$

12.10 Applications

In this chapter, we have used the same HRS data on mobility of elderly population for the period 1992-2004. We have considered 0= no difficulty, 1= difficulty in one or more of the five tasks.

Table 12.1 displays the distribution of elderly population by gender and mobility index for all the waves. It is observed that more females move from 0 to 1 compared to males. It appears from Table 12.2 as well, gender shows negative association with transition from 0 to 1 indicating females have higher transition to difficulty in mobility. Table 12.3 shows the stratified table for gender and mobility index by race (White and non-White). It appears that female among non-White races have much higher transition to difficulty in mobility compared to White females at older ages. To show the conditional and marginal models, we have chosen Models I and II (conditional) and Model in Table 12.2 (marginal). Gender appears to be significant in all the models, but for non-Whites the male-female discrimination is more prominent (Models I and II). The marginal model, presented in Table 12.2 for pooled data for race, indicates that the estimate is closer to that of Whites (Model II). To examine whether Whites have significant difference with non-Whites, a dummy variable for race (White=1, if race=White, White=0, if race=non-White) is included in the model (Model III). It indicates that race is an important variable in explaining the relationship between covariates and difficulty in mobility of elderly population. Hence, a marginal model is not appropriate. Model IV presents a further check of the relationship, between gender and race, by including the interaction term. It is revealed there is positive association between interaction of gender and race and difficulty in mobility.

Table 12.1. Distribution of Mobility Index by Gender among Elderly, 1992-2004

	Mobility Index		
	0	1+	Total
Female	15777	13368	29145
Male	17370	7692	25062
Total	33147	21060	54207

Table 12.2 Estimates of Parameters of Logit Model with Single Covariate for Mobility Index

Variables	Estimate	S.E.	Wald	p-value
Model I				
β_0	-0.166	0.012	198.663	0.000
Gender (β_1)	-0.649	0.018	1292.46	0.000
Model Chi-square	1317.605	(p=0.000)		
-2 Log Likelihood	71111.32			

Table 12.3. Distribution of Mobility Index by gender by race among elderly during 1992-2004

	Mobility Index					
	White			Other		
	0	1	Total	0	1	Total
Female	12810	9927	22737	2967	3441	6408
Male	14482	6145	20627	2888	1547	4435
Total	27292	16072	43364	5855	4988	10843

Table 12.4. Estimates of Parameters of Marginal and Conditional Logit Models for Mobility Index

Variables	Estimate	S.E.	Wald	p-value
Model I: Non White				
β_0	0.148	0.025	34.998	0.000
Gender (β_1)	-0.772	0.040	368.253	0.000
Model Chi-square	377.586	(p=0.000)		
-2 Log Likelihood	14584.61			
Model II: White				
β_0	-0.255	0.013	363.584	0.000
Gender (β_1)	-0.602	0.020	883.533	0.000
Model Chi-square	898.154	(p=0.000)		
-2 Log Likelihood	56280.76			
Model III:				
β_0	0.095	0.021	21.132	0.000
Gender (β_1)	-0.637	0.018	1238.835	0.000
White (β_2)	-0.335	0.022	233.256	0.000
Model Chi-square	1549.246	(p=0.000)		
-2 Log Likelihood	70879.68			
Model IV:				
β_0	0.148	0.025	34.998	.000
Gender (β_1)	-0.772	0.040	368.253	.000
White (β_2)	-0.403	0.028	201.565	.000
Gender * White (β_3)	0.170	0.045	14.256	.000
Model Chi-square	1563.561	(p=0.000)		
-2 Log Likelihood	70865.37			

Table 12.5 shows the conditional model based on consecutive follow-ups. Based on the outcomes in two consecutive follow-ups, we can fit two models for transition types 0-1 and 1-0. Taking gender as the covariate, we observe that gender is negatively associated with 0-1 transition but positively associated with 1-0.

Table 12.5. Estimates of Parameters of Conditional Model for Mobility Index Based on Consecutive Follow-up Data

Variables	Estimate	S.E.	t-value	p-value
$0 \rightarrow 1$				
β_0	-1.136	0.020	-56.450	0.000
Gender (β_1)	-0.515	0.030	-17.059	0.000
$1 \rightarrow 0$				
β_0	-1.320	0.024	-55.346	0.000
Gender (β_1)	0.271	0.038	7.091	0.000
Model Chi-square	15164.77	(0.000)		
LRT	16271.81	(0.000)		

The estimates of the PA GEE parameters are displayed in Table 12.6. The estimates are obtained for correlation structures independence, exchangeable, autoregressive and unstructured. It is observed that age and black race show positive association with outcome at different follow-ups while gender and White race show negative association. Similar findings are observed in Table 12.7 for the subject specific model. The model proposed by Azzalini (1994) also produces similar findings, positive association of age and black race and negative association of gender and White race with difficulty in mobility in old age (Table 12.8).

Table 12.6. Estimates of Parameters PA Model Using GEE for Mobility Index

Variables	Estimate	S.E.	Z-value	p-value
Independent Correlation				
Intercept	-3.0682	0.1713	-17.91	0.0001
Age	0.0496	0.0025	20.14	0.0001
Gender	-0.6407	0.0340	-18.84	0.0001
White	-0.1942	0.0881	-2.20	0.0275
Black	0.1911	0.0945	2.02	0.0432
Deviance	70102.87	Value/DF=	1.2934	
Pearson Chi-Square	54198.46	Value/DF=	0.9999	
Log Likelihood	-35051.43			
Exchangeable Correlation				
Intercept	-4.2999	0.1443	-29.80	0.0001
Age	0.0701	0.0019	36.21	0.0001
Gender	-0.5976	0.0337	-17.74	0.0001
White	-0.1826	0.0871	-2.10	0.0360
Black	0.2132	0.0934	2.28	0.0224
Deviance	70102.87	Value/DF=	1.2934	
Pearson Chi-Square	54198.46	Value/DF=	0.9999	
Log Likelihood	-35051.43			

Table 12.6. Continued

Variables	Estimate	S.E.	Z-value	p-value
Autoregressive Correlation				
Intercept	-3.7558	0.1197	-24.27	0.0001
Age	0.0608	0.0018	28.15	0.0001
Gender	-0.6089	0.0182	-18.22	0.0001
White	-0.1964	0.0485	-2.27	0.0230
Black	0.1855	0.0521	2.00	0.0453
Deviance	70102.87	Value/DF=	1.2934	
Pearson Chi-Square	54198.46	Value/DF=	0.9999	
Log Likelihood	-35051.43			
Unstructured Correlation				
Intercept	-4.1438	0.1436	-28.85	0.0001
Age	0.0674	0.0019	34.84	0.0001
Gender	-0.5950	0.0333	-17.86	0.0001
White	-0.1920	0.0861	-2.23	0.0258
Black	0.1977	0.0924	2.14	0.0323
Deviance	70102.87	Value/DF=	1.2934	
Pearson Chi-Square	54198.46	Value/DF=	0.9999	
Log Likelihood	-35051.43			

Table 12.7. Estimates of Parameters of Subject Specific Model Using GEE for Mobility Index

Variables	Estimate	S.E.	t-value	p-value
Intercept	-3.225	0.116	-27.870	0.000
Age	0.037	0.002	23.010	0.000
Gender	-0.434	0.023	-18.900	0.000
White	-0.129	0.060	-2.150	0.031
Black	0.130	0.064	2.030	0.042
SB2	0.602	0.021	28.890	0.000
-2 Log Likelihood	77443			

Table 12.8. Estimates of Parameters of Marginal Model (Azzalini) for Mobility Index

Variables	Estimate	S.E.	Z-value	p-value
Independent Correlation				
Intercept	-3.796419	0.169480	-22.400	0.0000
Age	0.061552	0.002573	23.919	0.0000
Gender	-0.621285	0.023785	-26.120	0.0000
White	-0.203624	0.065012	-3.132	0.0017
Black	0.168388	0.070104	2.402	0.0163
Lambda	2.533941	0.018008	140.708	0.0000
Log Likelihood	-19762.15			

12.11 SUMMARY

In this chapter the generalized linear model is further explored for logit models for contingency tables and then marginal and conditional approaches are described. One of the most extensively used techniques in the repeated measures analysis is the generalized estimating equations which is reviewed here for both the population averaged and subject specific approaches. Azzalini (1994) proposed a marginal model based on binary Markov chain. This chapter includes a comprehensive review of the method along with some of the limitations. We can also consider the regressive logistic regression model and other models proposed in previous chapter under conditional models. Hardin and Hilbe (2003) is suggested for a thorough understanding of estimating equations. Comparison of subject-specific and population-averaged models are displayed by Ten Have, Landis and Hartzel (1996), Hu et al. (1998) and Young et al. (2007). For a marginal model, collapsibility of logistic regression coefficients is discussed by Guo and Geng (1995). Lindsey and Lambert (1998) gave a very good account of the appropriateness of marginal models for repeated measurements. Bonney (1986, 1987) discussed regressive logistic models for dependent binary observations.

APPENDIX

COMPUTER PROGRAMS

A1. Data Files

We have used SAS or SPSS and customized SAS/IML software for application in this book. The customized SAS/IML software was used to estimate the parameters of covariate dependent Markov models and related tests. Before discussing how to run the programs, let us define the data file format used in the programs. For each follow-up, we have one record in the data file. The following table shows first 21 lines from the data file. First column is patient id, second column is follow-up number, third column is the dependent variable, and fourth onward are the independent variables or covariates. First row in the data file should be the variables names. It should be noted that the dependent variables should be coded as 0, 1 for binary dependent variable, 0, 1, 2 for dependent variable with 3 categories and so on. All records with missing value have to be removed. Our SAS/IML software will not handle the missing value.

Table A. Sample Data File for the SAS Program

CASEID	Wave	Mobility	AGE	GENDER	White	Black
1	1	0	54	1	1	0
1	2	1	56	1	1	0
2	1	1	57	0	1	0
2	2	1	59	0	1	0
2	3	1	62	0	1	0
2	4	1	63	0	1	0
2	5	1	65	0	1	0
3	1	0	56	1	1	0
3	2	0	58	1	1	0
3	3	0	60	1	1	0
3	4	0	62	1	1	0
3	5	0	64	1	1	0
3	6	0	66	1	1	0
3	7	1	68	1	1	0

Table A. (Continued)

CASEID	Wave	Mobility	AGE	GENDER	White	Black
4	1	0	54	0	1	0
4	2	0	55	0	1	0
4	3	1	57	0	1	0
4	4	0	59	0	1	0
4	5	0	61	0	1	0
4	6	0	63	0	1	0
4	7	0	65	0	1	0

In the sample data file above, Mobility is a binary dependent variable. This can also be a dependent variable with multiple categories. As mentioned earlier the data set we used in the book is a public domain data set. We can not provide the data set to any third party according to the data use condition. However, interested researchers can obtain the data set after acquiring necessary permission from the Health and Retirement Study site (http://hrsonline.isr.umich.edu/).

A2. SAS Programs for Chapter 2

Let us give some guidelines about how to use our SAS/IML customized program for parameter estimation of covariate dependent Markov models. All functions of SAS/IML customized program are stored in a file (mcfun.sas). This file has to be opened in SAS program editor, and then one has to select all the functions and run the program. It will be available for the current SAS session. Next step is to open the data file and call the SAS/IML function. The following SAS instructions show the opening of data file and running the customized SAS/IML program for all the applications used in Chapter 2.

```
PROC IMPORT OUT= WORK.mcdata
        DATAFILE= "g:\BOOKExample\BookChtwo.dat"
        DBMS=TAB REPLACE;
        GETNAMES=YES;
        DATAROW=2;
RUN;
```

The above SAS instruction opens the ASCII data file BookChtwo.dat from BOOKExample directory of G drive. It also names the data as WORK.mcdata SAS data set. As mentioned earlier the first row is the variable names in the data file. User's can use any SAS statements to read data files in different format.

Following SAS statements run our customized program.

```
PROC iml;
load module=udmload;
run udmload;
```

run mcmain(mcdata,2,1,1,0,1);

Statement **PROC iml** starts IML, second and third line load and run all the functions of our SAS/IML customized program. The last line invokes the main SAS/IML routines and estimates the parameters and related tests of Markov Model. We have to provide in total in six arguments to mcmain() function. First argument mcdata uses the SAS WORK.mcdata data opened in the PROC IMPORT statement. Second argument 2, is the number of categories (states) in the dependent variable for which the minimum is 2. The third argument 1 is the order of the Markov chain. This 1 is for the first order. For second order the third argument will be 2 and so on. Fourth argument is the maximum number of iterations which is 1 here. For examples in chapter 2 we need only pooled transition count, transition probability matrix and same for the consecutive follow-ups and the corresponding tests. We do not want any estimates of the parameters of the covariate dependent Markov model, which was the reason to set maximum iteration to 1.

It output produced are presented in Chapter 2 from Table 2.1 to Table 2.4. For computing the examples for the second order in Tables 2.5 to Table 2.7, we have to set the argument for order=2 to **run mcmain(mcdata,2,2,1,0,1);** We have to set the argument for order=3 for the third order and order=4 for all the examples for the fourth order Markov model.

SAS/IML Output for first order binary dependent variable using the following instructions:

PROC iml;
load module=udmload;
run udmload;
run mcmain(mcdata,2,**1**,**1**,**0**,**1**);

The results are displayed below:

```
                    Order of MC
                              1

                No. of States of MC
                                  2

            Diffrent Types of Transition

                    0                   0
                    0                   1
                    1                   0
                    1                   1

              Transition Count Matrix
                    0              1      Total
      0         22461           5621      28082
      1          3733          12636      16369

           Transition Probability Matrix
                    0              1      Total

      0         0.800          0.200      1.000
      1         0.228          0.772      1.000
```

```
                              MAXT          MAXT1
                   Time          1 ->          2

                  Transition Count & Probaility Matrix
          0             1       Total          0            1       Total

0    5239.00      1389.00    6628.00         0.79         0.21       1.00
1     645.00      1947.00    2592.00         0.25         0.75       1.00

                              MAXT          MAXT1
                   Time          2 ->          3

                  Transition Count & Probaility Matrix
          0             1       Total          0            1       Total

0    4340.00       988.00    5328.00         0.81         0.19       1.00
1     703.00      2196.00    2899.00         0.24         0.76       1.00

                              MAXT          MAXT1
                   Time          3 ->          4

                  Transition Count & Probaility Matrix
          0             1       Total          0            1       Total

0    3902.00       832.00    4734.00         0.82         0.18       1.00
1     678.00      2200.00    2878.00         0.24         0.76       1.00

                              MAXT          MAXT1
                   Time          4 ->          5

                  Transition Count & Probaility Matrix
          0             1       Total          0            1       Total

0    3420.00       826.00    4246.00         0.81         0.19       1.00
1     632.00      2093.00    2725.00         0.23         0.77       1.00

                              MAXT          MAXT1
                   Time          5 ->          6

                  Transition Count & Probaility Matrix
          0             1       Total          0            1       Total
0    2947.00       832.00    3779.00         0.78         0.22       1.00
1     542.00      2050.00    2592.00         0.21         0.79       1.00

                              MAXT          MAXT1
                   Time          6 ->          7

                  Transition Count & Probaility Matrix
          0             1       Total          0            1       Total

0    2613.00       754.00    3367.00         0.78         0.22       1.00
1     533.00      2150.00    2683.00         0.20         0.80       1.00

                         MC Statistical Inference
                              Test          d.f      p-value

                Chi-square= 14940.7708   2.000000   0.000000
                LRT       = 15927.3597   2.000000   0.000000

                           MC Stationary Test
                        T Chi-square          d.f     p-value
```

```
2.000000  25.313464   2.000000   0.000003
3.000000   4.166969   2.000000   0.124496
4.000000  11.152029   2.000000   0.003788
5.000000  25.091550   2.000000   0.000004
6.000000   2.065138   2.000000   0.356091

            Total Chi-square
      Chi-square        d.f     p-value
       67.789150  10.000000   0.000000


MC Stationary Test-Comparison with Polled  TPM
           T  Chi-square            d.f        p-value

2.000000      9.851460    2.000000     0.007257
3.000000     10.743862    2.000000     0.004645
4.000000     19.119211    2.000000     0.000071
5.000000      1.077307    2.000000     0.583533
6.000000     14.612629    2.000000     0.000671
7.000000     25.153943    2.000000     0.000003

            Total Chi-square
      Chi-square        d.f     p-value
       80.558411  12.000000   0.000000

            Iteration Number
                          1

            MC Estimates for Transition Type 01
           Coeff.   Std. err.     t-value    p-value    .95CI LL     .95CI UL

Const     -1.853549   0.155544 -11.916558   0.000000  -2.158416  -1.548683
r1agey_b   0.010997   0.002607   4.218348   0.000025   0.005887   0.016106

            MC Estimates for Transition Type 10
           Coeff.   Std. err.     t-value    p-value    .95CI LL     .95CI UL

Const     -0.316425   0.210371  -1.504130   0.132548  -0.728751   0.095902
r1agey_b  -0.012770   0.003473  -3.676854   0.000236  -0.019578  -0.005963


                         MC Model Test

                              Test          d.f       p-value
U(B0)*inv(I(B0))*U(B0)  14972.0845    4.000000    0.000000
U(B)*inv(I(B))*U(B)     14972.0845    4.000000    0.000000
(BH-B0)*I(BH)*(BH-B0)   14972.0845    4.000000    0.000000
(BH-B0)*I(B0)*(BH-B0)   14972.0845    4.000000    0.000000
Sum (Zi-square)         175.580480    4.000000    0.000000
LRT                       0.000000    4.000000    1.000000
AIC                     61630.1706
BIC                     61658.9132

Function has not converged..Try by increasing max iteration
```

In Chapter 2, Table 2.1 was prepared from "Transition Count Matrix and Transition Probability Matrix" of above output. The test statistic for the first order Markov chain in Table 2.2 is taken from "MC Statistical Inference" of the above output. After the pooled transition counts and the transition probabilities are computed, the output shows the consecutive transition counts and the probabilities in the above output which are presented in Table 2.3. The "MC Stationarity Tests" are based on the consecutive follow-ups and "MC Stationarity Test comparison with the pooled TPM" are presented in Table 2.4. In addition, it also shows the total chi-square which is the sum of chi-squares for all follow-ups. Then it shows the estimate of the parameters (constant and the coefficients of age=r1agey_b) of the Markov model and test related to the model fit, which is not used for chapter 2. If the message appears at the end "Function has not converged…then try by increasing max iteration" it tells us that the estimate did not converge because we used the maximum iteration as 1 which we set for the fourth argument.

A3. SAS and SPSS programs for examples in Chapter 3

The following SAS statements open the data file for Chapter 3 examples.

```
PROC IMPORT OUT= WORK.Mobility
        DATAFILE= "g:\BOOKExample\BookChthree.dat"
        DBMS=TAB REPLACE;
        GETNAMES=YES;
        DATAROW=2;
RUN;
```

The example presented in Table 3.1 in Chapter 3 is based on only from the 1992 survey data. Following SAS statements create a new data set Mobility1 by selecting only the records from the first wave (1992 survey).

```
DATA Mobility1;
   SET Mobility;
   WHERE WAVE=1;
RUN;
```

To run the logistic regression for a single covariate age (r1agey_b) which is presented in Table 3.1 in Chapter 3, we have used the following SAS statements. The dependent variable used in model statement r1mobil is binary (0, 1). It should be noted that we have not presented all the results in the table from the SAS output.

```
PROC LOGISTIC DATA=Mobility1 DESCENDING;
        MODEL r1mobil = r1agey_b/ SCALE= D CLPARM=WALD CLODDS=PL
   RSQUARE   OUTROC=ROC1;
RUN;
```

The following SAS statements run the logistic regression procedure for three more covariates as compared to the previous SAS statements. The results are presented in Table 3.2.

```
PROC LOGISTIC DATA=Mobility1 DESCENDING;
MODEL r1mobil = r1agey_b ragender rawhca rablafa
        / SCALE= D CLPARM=WALD CLODDS=PL RSQUARE OUTROC=ROC1;
RUN
```

The multinomial logistic regression estimates presented in Table 3.3, can be estimated using the SAS CATMOD procedure. The dependent variable MOBILS3 has three categories (0,1,2). However, we used the following SPSS syntax for the results presented in Table 3.3.

```
USE ALL.
COMPUTE filter_$=(WAVE = 1).
VARIABLE LABEL filter_$ 'WAVE = 1 (FILTER)'.
VALUE LABELS filter_$  0 'Not Selected' 1 'Selected'.
FORMAT filter_$ (f1.0).
FILTER BY filter_$.
EXECUTE .
```

Above SPSS syntax selects the cases from wave 1 (1992 survey) and the following SPSS syntax is used to run the multinomial logistic regression estimates presented in Table 3.3. For details, please consult SPSS manual. The same can be run from the SPSS windows menu.

```
NOMREG
  MOBILS3  (BASE=FIRST ORDER=ASCENDING) WITH r1agey_b ragender rawhca
rablafa
  /CRITERIA   CIN(95)   DELTA(0)   MXITER(100)   MXSTEP(5)   CHKSEP(20)
LCONVERGE(0)
  PCONVERGE(0.000001) SINGULAR(0.00000001)
  /MODEL
  /STEPWISE   =   PIN(.05)   POUT(0.1)   MINEFFECT(0)   RULE(SINGLE)
ENTRYMETHOD(LR)
  REMOVALMETHOD(LR)
  /INTERCEPT =INCLUDE
  /PRINT = PARAMETER SUMMARY LRT CPS STEP MFI .
```

The results presented in Table 3.4, are obtained from the SAS output by using the following SAS statements. First, the DATA procedure is used to create a new data set by selecting the record from the First 2 waves (1992 & 1994 survey). The PROC LOGISTIC is used to run the logistic regression procedure.

```
DATA Mobility2;
   SET Mobility;
   WHERE WAVE<=2;
RUN;

PROC LOGISTIC DATA=Mobility2 DESCENDING;
MODEL r1mobil = r1agey_b ragender rawhca rablafa
        / SCALE= D CLPARM=WALD CLODDS=PL RSQUARE OUTROC=ROC1;
```

RUN

The multinomial logistic regression estimates are presented in Table 3.5 and the results are obtained from the SPSS output using the following SPSS syntax.

```
USE ALL.
COMPUTE filter_$=(WAVE <= 2).
VARIABLE LABEL filter_$ 'WAVE <= 2 (FILTER)'.
VALUE LABELS filter_$  0 'Not Selected' 1 'Selected'.
FORMAT filter_$ (f1.0).
FILTER BY filter_$.
EXECUTE .
```

Above SPSS syntax selects the cases from wave 1 and wave 2 (1992 & 1994 survey) and the following SPSS syntax is used to run the multinomial logistic regression presented in Table 3.5.

```
NOMREG
  MOBILS3  (BASE=FIRST ORDER=ASCENDING) WITH r1agey_b ragender rawhca
rablafa
  /CRITERIA   CIN(95)   DELTA(0)   MXITER(100)   MXSTEP(5)   CHKSEP(20)
LCONVERGE(0)
  PCONVERGE(0.000001) SINGULAR(0.00000001)
  /MODEL
  /STEPWISE   =   PIN(.05)   POUT(0.1)   MINEFFECT(0)   RULE(SINGLE)
ENTRYMETHOD(LR)
  REMOVALMETHOD(LR)
  /INTERCEPT =INCLUDE
  /PRINT = PARAMETER SUMMARY LRT CPS STEP MFI .
```

The results presented in Table 3.6 provide the estimates from the multiple logistic regression The previous outcome is used as a covariate. Following SPSS syntax were used. First we created Y1 by using the lag function using the following SPSS syntax.

```
COMPUTE Y1 = lag(r1mobil,1) .
EXECUTE .
```

Following SPSS syntax selects records from wave 2 (1994 survey) only.

```
USE ALL.
COMPUTE filter_$=(WAVE = 2).
VARIABLE LABEL filter_$ 'WAVE = 2 (FILTER)'.
VALUE LABELS filter_$  0 'Not Selected' 1 'Selected'.
FORMAT filter_$ (f1.0).
FILTER BY filter_$.
EXECUTE .
```

Finally, the following SPSS syntax is used to run the logistic procedure from which we obtained the results presented in Table 3.6.

```
LOGISTIC REGRESSION r1mobil
 /METHOD = ENTER r1agey_b ragender rawhca rablafa Y1
 /PRINT = CI(95)
 /CRITERIA = PIN(.05) POUT(.10) ITERATE(20) CUT(.5) .
```

A4. SAS program for examples in Chapter 4 to Chapter 6

The following SAS instruction opens the ASCII data file BookCh4-6.dat from BOOKExample directory of G drive. It names the data as WORK.mcdata SAS data set. As mentioned earlier the first row is the variables names in the data file. In this file the dependent variable is binary. Also one should follow same file format as explained in the sample data file format.

```
PROC IMPORT OUT= WORK.Mcdata
        DATAFILE= "g:\BOOKExample\BookCh4-6.dat"
        DBMS=TAB REPLACE;
        GETNAMES=YES;
        DATAROW=2;
RUN;
```

We have explained in the beginning about how to run a program and about the data file format of our customized SAS/IML program. We have used the following statements to obtain the results presented in Chapters 4 to 6. The fourth argument which is for maximum iteration can be changed to a higher value (e.g., 31). If the function does not converge then the maximum iteration value needs to be increased to a higher value. If there are not enough cell frequencies for some transition types then the function will not converge at all. This can be checked by going through the "Transition Count Matrix".

For Chapter 4 we have used the following arguments to our customized program.

```
PROC iml;
load module=udmload;
run udmload;
run mcmain(mcdata,2,1,31,0,1);
```

For Chapter 5, the third parameter in mcmain() function which is for order of the Markov chain is set to 2. For the third and the fourth order models presented in Chapter 6, we set the third argument to 3 and 4, respectively, in the mcmain() function.

A5. SAS Programs for Chapter 7 and Chapter 8

Chapter 7 presents the multi-state Markov model. We need to open a new data file where the dependent variable has three categories (0,1,2). Following SAS DATA procedure open the data file.

```
PROC IMPORT OUT= WORK.Mcdata
        DATAFILE= "g:\BOOKExample\BookCh7-8.dat"
        DBMS=TAB REPLACE;
        GETNAMES=YES;
        DATAROW=2;
RUN;
```

To obtain the results presented in Chapter 7, we need to set the arguments for states and order to 3 and 1, respectively, for 3 states 1st order covaraite dependent Markov model.

```
PROC iml;
load module=udmload;
run udmload;
run mcmain(mcdata,3,1,31,0,1);
```

For 3 states 2nd order Markov model we need to set the third argument as 2.

```
PROC iml;
load module=udmload;
run udmload;
run mcmain(mcdata,3,2,31,0,1);
```

A6. SAS Programs for Chapter 9

The computer program for the parameter estimation of the model proposed in Chapter 9 is displayed here. The estimation of the model parameters proposed in this chapter is complicated and tedious. In addition it needs to be done in several phases and a large amount of pre processing for data preparation is needed. This model used the follow-up data for those cases whose outcome variable of interest shall start only with state 0 at Wave 1 (follow-up1). Depending on the number of states and number of follow-ups lots of comparison needs to be done to create the data file for transition, reverse transition, repeated transitions and cases without change of states throughout the follow-up times. Table B shows the possible comparisons that are needed to create the data for transitions, reverse transitions, repeated transitions and cases without changes of state for a study with four follow-ups and two states (0, 1) only.

Table B. Possible comparison for 2 states with 4 follow-ups for model in chapter 9

		Follow-ups			
Transition Type	States	1	2	3	4
<u>Transition:</u> No changes	0	0	0	0	0
	0	0	99	.	.
	0	0	0	99	.
	0	0	0	0	99
Transition from state 0 to state 1	1	0	1	.	.
	1	0	0	1	.
	1	0	0	0	1
<u>Reverse Transition:</u> No changes after Making a transition	0	0	1	1	1
	0	0	0	1	1
	0	0	1	99	.
	0	0	0	1	99
Reverse Transition 0 to 1 then 0	1	0	1	0	.
	1	0	1	1	0
<u>Repeated Transition:</u> No changes after Making a reverse transition	0	0	1	0	0
Repeated Transition 0 to 1 to 0 then 1	1	0	1	0	1

Note: 99's are missing values which are considered as sate 0 (censored) that is no change.

With the increasing number of follow-ups and states, the above combinations will increase accordingly and identifying the cases for transitions, reverse transitions, repeated transitions and cases without change of states and the corresponding covariates will be tedious and erroneous. We wrote a SAS/IML function trrmain() to make this processing automated. Our SAS/IML function will read a FLAT file and create the data file with transition, reverse transition, repeated transitions and cases without changes of state and the corresponding covariates. The following SAS statements read the ASCII data file "BookCh9FLAT.dat" from the specified directory. This is a flat ASCII file. First column is the CASE ID, from 2nd to 5th column is the outcome variable for the follow-ups 1 to 4. The outcome variable should be coded as 0, 1, 2 and so on according to the number of states. From columns 6 to column 9 should be the first covariates for four follow-ups respectively. The same pattern has to be followed for other covariates. It should be mentioned that this flat file consists of only those cases which are in states 0 at first follow-up. Missing values should be coded as 99 for outcome variables only. However, covariates may contain missing values. The first row in the data file shall be the variable names.

```
PROC IMPORT OUT= WORK.trdata
    DATAFILE= "g:\BOOKExample\BookCh9FLAT.dat"
    DBMS=TAB REPLACE;
    GETNAMES=YES;
    DATAROW=2;
RUN;
```

To use our function trrmain(), we need to open the trrmain.SAS file in editor window and run the whole program. Then we need to submit the following SAS statements to RUN our function.

```
proc iml;
load module=trrload;
run trrload;
run trrmain(trdata,2,7,0,{0},3);
```

The first argument in trrmain() is trdata which is the SAS data loaded by the PROC IMPORT statement above. The second argument 2 is the number of states (0, 1). The third argument 7 is the total number of follow-ups in our data file. The fourth argument is for missing values of the outcome variable. We have used 0 as a state of no change; it can be coded as a separate state by setting 99 instead of 0. The fifth argument in the {0} should be fixed. We work to identify this argument for the absorbing states. The last argument 3 is for the transition types. For the transition type column in the resulting data file, our function will set value 1 for transitions, 2 for reverse transitions, and 3 for repeated transitions. If we set the last argument to 2 then it will create data for transitions and reverse transitions only and for 1 it will be for transition only. Our function will create a data set called "Fdatres" in the SAS WORK library. This data set can be saved as a separate file using the standard SAS statements. The first column in the resulting data will be CASEID, the second column will be Trantype and the values for second column will be 1,2,3 for transitions, reverse transition, and repeated transitions, respectively. The third column is TranCode, we do not need to use it right now. The fourth column is the code for states as shown in TABLE A. The fifth and the next onward will be the covariates according to the number of covariates in the data file. It should be noted that there will be only one covariate irrespective of number of follow-ups in the original data file.

Then we can use the SAS or the SPSS to run the binomial or the multinomial logistic regression models by selecting transitions, reverses transitions, and repeated transitions separately. StateCode shall be used as dependent variable. We have used the SPSS to produce the Tables 9.1 and 9.2. Our program can be used for any number of states and follow-ups. However, if there is any absorbing state then we need further processing and some modifications and additions are needed to our program which we have decided not to present in this edition of the book. Tables 9.3 and 9.4 show the same applications for three states using death as an absorbing state.

A7. SPSS and SAS Programs for Chapter 10

In this chapter we have used the data from the Maternal Morbidity Survey described in Chapter 1. The results presented in Table 10.5 are based on the estimates of the multiple logistic regression parameters for the follow-up 1 to the follow-up 6 separately. However, the previous outcome is used as a covariate. The following SPSS syntax provides an outline of how we can perform the analysis. The model for the follow-up 1 in Table 10.5 can be done using the following SPSS syntax.

```
USE ALL.
COMPUTE filter_$=(FUP = 1).
VARIABLE LABEL filter_$ 'FUP = 1 (FILTER)'.
VALUE LABELS filter_$  0 'Not Selected' 1 'Selected'.
FORMAT filter_$ (f1.0).
FILTER BY filter_$.
EXECUTE .
```

Above syntax will select the first follow-ups only of all the cases and the following syntax will run the logistic regression where compli is a binary dependent variable and v50 (wanted pregnancy), v12 (gainful employment), agem (age at marriage), edu (education), and npreg (number of pregnancies) are independent variables.

```
LOGISTIC REGRESSION  compli
 /METHOD = ENTER v50 v12 agem edu npreg
 /PRINT = CI(95)  /CRITERIA = PIN(.05) POUT(.10) ITERATE(20) CUT(.5) .
```

The logistic regression estimates from the follow-up 2 to the follow-up 6, presented in table 10.5, are obtained on the basis of the following steps.

If we need to create a covariate by taking the previous follow-ups outcome then the following SPSS syntax will do the needful.

```
COMPUTE Y1 = lag(compli,1) .
EXECUTE .

USE ALL.
COMPUTE filter_$=(FUP = 2).
VARIABLE LABEL filter_$ 'FUP = 2 (FILTER)'.
VALUE LABELS filter_$  0 'Not Selected' 1 'Selected'.
FORMAT filter_$ (f1.0).
FILTER BY filter_$.
EXECUTE .
```

Above syntax will select all cases from the second follow-up only and the following syntax will run the logistic regression. Here compli is a binary dependent variable and v50 (wanted pregnancy), v12 (gainful employment), agem (age at marriage), edu (education), npreg (number of pregnancies), and Y1 (outcome at previous follow-up) are independent variables.

```
LOGISTIC REGRESSION  compli
 /METHOD = ENTER v50 v12 agem edu npreg Y1
 /PRINT = CI(95)  /CRITERIA = PIN(.05) POUT(.10) ITERATE(20) CUT(.5) .
```

We can repeat the above two steps just by selecting follow-up 3 and so on for the rest of the results presented in Table 10.5.

The results presented in Table 10.6 can be run similar way as we have done for Table 10.5 using the SPSS syntax in the previous steps for follow-up 1 and follow-up 2. However, in the model for follow-up 3 in Table 10.6, we have two extra covariates those were created by taking previous two follow-ups of out outcome variable as covariates. To do so we need to create two lag variables from the outcome variables of follow-up 1 and the follow-up 2. This can be done using the following SPSS syntax.

```
COMPUTE Y1 = lag(compli,1) .
EXECUTE .

COMPUTE Y2 = lag(compli,2) .
EXECUTE .
```

Following syntax will select all cases from the third follow-up only

```
USE ALL.
COMPUTE filter_$=(FUP = 3).
VARIABLE LABEL filter_$ 'FUP = 3 (FILTER)'.
VALUE LABELS filter_$  0 'Not Selected' 1 'Selected'.
FORMAT filter_$ (f1.0).
FILTER BY filter_$.
EXECUTE .
```

To run the logistic regression we can use the following syntax.

```
LOGISTIC REGRESSION  compli
 /METHOD = ENTER v50 v12 agem edu npreg Y1 Y2
 /PRINT = CI(95)  /CRITERIA = PIN(.05) POUT(.10) ITERATE(20) CUT(.5) .
```

We can repeat the above three steps for creating covariates from the outcome variables for all the previous follow-ups and by selecting the follow-ups for which we need to run the logistic regression.

The results presented in Tables 10.7 and 10.8 are done using our SAS/IML function mcmain(mcdata,2,1,31,0,1) described in the beginning and for example in Chapter 4. Only exception for this example is "LR test for the Waiting Time" at the last row of Table 10.7. We need to run our SAS/IML function twice, first by taking all the covariates except the waiting time, second by rerunning the SAS/IML function by including waiting time. We need to do this to calculate the "LR test for the Waiting Time".

A8. SAS Programs for Chapter 11

The results presented in Table 11.1 are the transition count matrix copied from Table 2.1 which is for the first order Markov model for 2 states and we used the count data to estimate the parameters of the log linear model. We have used the following SAS statements to obtain the parameter estimates for Model I in Table 11.2.

```
DATA mob1;
INPUT y1 y2 count @@;
DATALINES;
2 2 22461 2 1 5621
1 2 3733  1 1 12636;
RUN;
```

Above data statements are used to read the count data presented in Table 11.1. To estimate the parameters of the log linear model assuming the Poisson distribution and the log link we have used the PROC GENMOD as follows.

```
PROC GENMOD DATA=mob1 DESC ORDER= DATA;
 CLASS y1 y2;
 MODEL count= y1 y2 /DIST=poi LINK=log lrci type3;
 RUN;
```

The results presented under Model II in Table 11.2 have the estimates for interaction term compared to results presented under Model I. We have used the following SAS statements.

```
PROC GENMOD DATA=mob1 DESC ORDER= DATA;
 CLASS y1 y2;
 MODEL count= y1 y2 y1*y2/DIST=poi LINK=log lrci type3;
 RUN;
```

The results presented in Table 11.3 are the transition count matrix copied from Table 2.5 which is for the second order Markov model for 2 states and we have used the count data to estimate the parameters of the log linear model. We have used the following SAS statements to compute the parameter estimates under Model IV in Table 11.4.

```
DATA mob3;
INPUT y3 y2 y1 count @@;
DATALINES;
2 2 2 15687  2 2 1 2833 2 1 2 1825  2 1 1 2589
1 2 2 1535   1 2 1 1399 1 1 2 1263  1 1 1 8098;
RUN;
```

Above data statements are used to read the count data presented in Table 11.3. To estimate the parameters of the log linear model assuming the Poisson distribution and the log link we have used the PROC GENMOD as follows.

```
PROC GENMOD DATA=mob3 DESC ORDER= DATA;
 CLASS y1 y2 y3;
 MODEL count= y1 y2 y3/DIST=poi LINK=log lrci type3;
RUN;
```

The following SAS statements are used to obtain the parameter estimates under Model V in Table 11.4 using the same data (mob3). This model includes all two-way interactions in addition to the main effects.

```
PROC GENMOD DATA=mob3 DESC ORDER= DATA;
 CLASS y1 y2 y3;
 MODEL count= y1 y2 y3 y1*y2 y1*y3 y2*y3/DIST=poi LINK=log
lrci ype3;
RUN;
```

The following SAS statements are used for Model VI in Table 11.4 which includes all the main effects and the higher order interactions from the same data (mob3).

```
PROC GENMOD DATA=mob3 DESC ORDER= DATA;
 CLASS y1 y2 y3;
 MODEL count= y1 y2 y3 y1*y2 y1*y3 y2*y3 y1*y2*y3/DIST=poi
LINK=log lrci ype3;
RUN;
```

The results presented in Table 11.5 are the estimates of the log linear model for the binomial distribution and the logit link. However we have used the previous outcome as covariates too. We have already shown how we can create covariates from the previous waves or follow-ups. We have used the file created for Chapter 10. The following PROC IMPORT statement opens the Chap11BinY1.dat ASCII file which includes Y1 as covariates were created from WAVE I of the outcome variables. It should be noted that this data file contains cases from WAVE II to WAVE VII.

```
PROC IMPORT OUT= WORK.mcdata
      DATAFILE= "g:\BOOKExample\Chap11BinY1.dat"
      DBMS=TAB REPLACE;
      GETNAMES=YES;
      DATAROW=2;
RUN;
```

The following SAS GENMOD procedure is used to compute the results presented under Model I in Table 11.5.

```
PROC GENMOD DATA=Mcdata DESC ORDER= DATA;
 CLASS y1;
 MODEL r1mobil= y1/DIST=bin LINK=logit lrci type3;
RUN;
```

The following PROC IMPORT statement opens the Chap11BinY1Y2.dat ASCII file which includes Y1 and Y2 as covariates which are created from WAVE I and WAVE II of the outcome variables. It should be noted that this data file contains cases from WAVE III to WAVE VII.

```
PROC IMPORT OUT= WORK.mcdata
       DATAFILE= "g:\BOOKExample\Chap11BinY1Y2.dat"
       DBMS=TAB REPLACE;
       GETNAMES=YES;
       DATAROW=2;
RUN;
```

The following SAS GENMOD procedure is used to compute the results presented under Model II in Table 11.5.

```
PROC GENMOD DATA=Mcdata DESC ORDER= DATA;
 CLASS y1 y2;
 MODEL r1mobil= y1 y2/DIST=bin LINK=logit lrci type3;
RUN;
```

For Model III which includes interaction term we used following SAS statements.

```
PROC GENMOD DATA=Mcdata DESC ORDER= DATA;
 CLASS y1 y2;
 MODEL r1mobil= y1 y2 y1*y2/DIST=bin LINK=logit lrci type3;
RUN;
```

The following PROC IMPORT statement opens the Chap11BinY1Y2Y3.dat ASCII file which includes Y1, Y2, and Y3 as covariates which are created from WAVE I, WAVE II, and WAVE III of the outcome variables. It should be noted that this data file contains cases from WAVE IV and WAVE VII only because of the creation of lag variables from previous three waves.

```
PROC IMPORT OUT= WORK.mcdata
       DATAFILE= "g:\BOOKExample\Chap11BinY1Y2Y23.dat"
       DBMS=TAB REPLACE;
       GETNAMES=YES;
       DATAROW=2;
RUN;
```

For Model IV which includes main effects only, we have used the following SAS statements.

```
PROC GENMOD data=Mcdata DESC ORDER= DATA;
 CLASS y1 y2 y3;
 MODEL r1mobil= y1 y2 y3/DIST=bin LINK=logit lrci type3;
RUN;
```

For Model V which includes main effects and all two-way interaction terms, we can use the following SAS statements.

```
PROC GENMOD DATA=Mcdata DESC ORDER= DATA;
 CLASS y1 y2 y3;
 MODEL r1mobil= y1 y2 y3 y1*y2 Y1*y3 Y2*y3/DIST=bin LINK=logit
lrci type3;
RUN;
```

For Model VI which includes main effects, all two-way, and three-way interaction terms, we can employ the following SAS statements.

```
PROC GENMOD DATA=Mcdata DESC ORDER= DATA;
 CLASS y1 y2 y3;
 MODEL r1mobil= y1 y2 y3 y1*y2 Y1*y3 Y2*y3 y1*y2*y3
 /DIST=bin LINK=logit lrci type3;
RUN;
```

A9. SAS and SPSS Programs for Chapter 12

In the applications, shown from Tables 12.1 to 12.4 in Chapter 12, we have used the SPSS. The results presented in Table 12.1 are a simple cross table of gender by outcome variable (Mobility Index). The binomial logistic regression is fitted using the SPSS by taking gender as the only independent variable for Table 12.2. We are not repeating the SPSS syntax as we have already explained before. Again the results in Table 12.3 are a simple cross table of gender by two dummy variables of races namely White and Others.

The logistic regression estimates presented under Model I in Table 12.4 are only for Non-white respondents by taking gender as the only covariate and Model II is the same for white respondents. Model III is displayed for all the respondents. However, in addition to gender, we have added white race as a covariate too. Model IV is for covariates gender and white race and the interaction between gender and white . The SPSS binomial logistic regression procedure was used for the above applications.

The example presented in table 12.5 is based on a first order Markov model for binary outcome variable (Mobility Index) by taking gender as only covariate. We have used the SAS/IML function for covariate dependent Markov model as shown before.

The GEE-PA analysis is presented for different correlation structures in table 12.6 using the SAS for this analysis. Following SAS statements are used to open the data file and GEE-PA analysis.

```
PROC IMPORT OUT= WORK.mcdata
         DATAFILE= "g:\BOOKExample\BookCh12Bin.dat"
        DBMS=TAB REPLACE;
        GETNAMES=YES;
        DATAROW=2;
RUN;
```

Following statements are used to estimate the parameters of the GEE-PA for independent correlation structure.

```
PROC GENMOD DATA=Mcdata DESCEND;
 CLASS hhidpn;
 MODEL r1mobil= r1agey_b ragender rawhca rablafa/dist=bin
link=logit;
 REPEATED subject=hhidpn / corr=IND;
RUN;
```

Similarly, following statements are used to estimate the parameters of the GEE-PA for exchangeable correlation structure.

```
PROC GENMOD DATA=Mcdata DESCEND;
 CLASS hhidpn;
 MODEL r1mobil= r1agey_b ragender rawhca rablafa/dist=bin
link=logit;
 REPEATED subject=hhidpn / corr=EXCH;
RUN;
```

To estimate the parameters of the GEE-PA for the autoregressive correlation structure, we can use the statements displayed below:

```
PROC GENMOD DATA=Mcdata DESCEND;
 CLASS hhidpn;
 MODEL r1mobil= r1agey_b ragender rawhca rablafa/dist=bin
link=logit;
 REPEATED subject=hhidpn / corr=AR;
RUN;
```

For the unstructured correlation, the parameters of the GEE-PA are estimated as follows:

```
PROC GENMOD DATA=Mcdata DESCEND;
 CLASS hhidpn;
 MODEL r1mobil= r1agey_b ragender rawhca rablafa/dist=bin
link=logit;
 REPEATED subject=hhidpn / corr=UNST;
RUN;
```

The GEE-SS analysis is presented in Table 12.7 where we have used the SAS NLMIXED procedure. The Poisson distribution is assumed for the dependent variable. The distribution considered for the random effects is normal. Following SAS statements were used for analysis:

```
PROC NLMIXED DATA=Mcdata;
PARMS b0=0 b1=0 b2=0 b3=0 b4=0 sb2=1;
eta = b0+b1*r1agey_b + b2*ragender + b3*rawhca + b4*rablafa +
bi;
```

```
f = exp(eta);
MODEL r1mobil ~ poisson(f);
RANDOM bi ~ normal(0,sb2) subject=hhidpn;
RUN;
```

The example for the marginal model proposed by Azzalini is presented in Table 12.8. We have used our own SAS/IML function marmain()for parameter estimation of marginal model. Before we explain how to run our marmain() function let us explain how the data file format will be used by this function. The data format will be same as shown in Table A except for the fourth column which is the number of follow-ups for each case (see Table C). This column in the data file can be created either by the SPSS or by the SAS. It is noteworthy that the data file should not contain any missing values and there should be at least 2 follow-ups for a respondent. To use our SAS/IML marmain() function, one has to first open the marmain.SAS file in the program editor window and run the whole program. The next step is to open the data file using the following PROC IMPORT procedure.

```
PROC IMPORT OUT= WORK.trdata
    DATAFILE= "g:\BOOKExample\BOOKxampleChap12MF.dat"
    DBMS=TAB REPLACE;
    GETNAMES=YES;
    DATAROW=2;
RUN;
```

To invoke our marmain() function run the following SAS statements, first argument in marmain() is the data set opened by the PROC IMPORT statement and the second argument is the maximum number of iterations which can be changed.

```
proc iml;
load module=marload;
run marload;
run marmain(trdata,71);
```

Table C. Sample Data File for the SAS Program for example in chapter 12

CASEID	Wave	Mobility	MF	AGE	GENDER	White	Black
1	1	0	2	54	1	1	0
1	2	1	2	56	1	1	0
2	1	1	5	57	0	1	0
2	2	1	5	59	0	1	0
2	3	1	5	62	0	1	0
2	4	1	5	63	0	1	0
2	5	1	5	65	0	1	0
3	1	0	7	56	1	1	0
3	2	0	7	58	1	1	0
3	3	0	7	60	1	1	0
3	4	0	7	62	1	1	0

Table C. Continued

CASEID	Wave	Mobility	MF	AGE	GENDER	White	Black
3	5	0	7	64	1	1	0
3	6	0	7	66	1	1	0
3	7	1	7	68	1	1	0
4	1	0	7	54	0	1	0
4	2	0	7	55	0	1	0
4	3	1	7	57	0	1	0
4	4	0	7	59	0	1	0
4	5	0	7	61	0	1	0
4	6	0	7	63	0	1	0
4	7	0	7	65	0	1	0

REFERENCES

Agresti A. (2002). Categorical Data Analysis (Second Edition). New York, Wiley.

Akhter H.A., Chowdhury M.E.E.K., and Sen A. (1996). A cross-sectional study on maternal morbidity in Bangladesh. *Bangladesh Institute of Research for Health and Technologies* (BIRPERHT). Dhaka.

Albert P.S. (1994). A Markov Model for Sequence of Ordinal Data from a relapsing-remitting disease. *Biometrics,* 50: 51-60.

Albert P.S. and Waclawiw M.A. (1998). Two State Markov Chain for Heterogeneous Transitional Data: A Quasilikelihood Approach. *Statistics in Medicine 1998,* 17:1481-1493.

Anderson T.W and Goodman L. (1957). Statistical Inference about Markov Chains. *Annals of Mathematical Statistics,* 28: 89-110.

Avery P.J. (2002). Fitting Interconnected Markov Chain Models: DNA Sequences and Test Cricket Matches. *The Statistician* 51: 267-278.

Avery P.J. and Henderson D.A. (1999). Fitting Markov Chain Models tp Discrete State Series Such as DNA Sequences. *Applied Statistics* 48: 53-61.

Azzalini A. (1994). Logistic Regression for Autocorrelated Data with Application to Repeated Measures. *Biometrika* 81: 767-775.

Bartlett M.S. (1951). The frequency Goodness of Fit for Probability Chains. *Proc. Camb. Phil. Soc.* 47: 86-95.

Berchtold A and Raftery A.E. (2002). The Mixture Transition Distribution Model for High-Order Markov Chains and Non-Gaussian Time Series. *Statistical Science*, 17:328-356.

Bhat U.N. (1971). Elements of Applied Stochastic Processes. Wiley: New York.

Billingsley P. (1961). Statistical Methods in Markov Chains. *The Annals of Mathematical Statistics* 32: 12-40.

Bonney G.E. (1986). Regressive Logistic Models for Familial Disease and Other Binary Traits. *Biometrics,* 42:611-625.

Bonney G.E. (1987). Logistic regression for dependent binary observations. *Biometrics,* 43:951-973.

Carey V.C., Zeger S. L. and Diggle, P. J. (1993). Modelling multivariate binary data with alternating logistic regressions. *Biometrika* 80:517-26.

Chatfield C. (1973). Statistical Inference Regarding Markov Chain Models. *Applied Statistics* 22: 7-20.

De Stavola B.L. (1988). Testing Departures from Time Homogeneity in Multistate Markov Processes. *Applied Statistics* 37: 242-250.

Diggle P.J., Heagerty P.J., Liang K.Y., and Zeger S.L. (2002). Analysis of Longitudinal Data (Second Edition). Oxford University Press, Oxford.

Duncan G.T. and Lin L.G. (1972). Inference for Markov Chains Having Stochastic Entry and Exit. *Journal of the American Statistical Association* 67: 761-767.

Fitzmaurice G.M. and Lipsitz S.R. (1995). A Model for Binary Time Series Data with Serial Odds Ratio Patterns. *Applied Statistics* 44: 51-61.

Gold R.Z. (1963). Tests Auxiliary to #2 Tests in a Markov Chain. *The Annals of Mathematical Statistics* 34: 56-74.

Good I.J. (1955). The Likelihood Ratio Test for Markoff Chains. *Biometrika* 42: 531-533.

Goodman L.A. (1958a). Simplified Runs Tests and Likelihood Ratio Tests for Markoff Chains. *Biometrika* 45: 181-197.

Goodman L.A. (1958b). Exact Probabilities and Asymptotic Relationships for Some Statistics from m-th Order Markov Chains. *The Annals of Mathematical Statistics* 29: 476-490.

Goodman L.A. (1958c). Asymptotic Distributions of "Psi-Squared" Goodness of Fit Criteria for m-th Order Markov Chains. *The Annals of Mathematical Statistics* 29: 1123-1133.

Guo J. and Geng, Z. (1995). Collapsibility of Logistic Regression Coefficients. *Journal of the Royal Statistical Society. Series B (Methodological)*, 57:263-267.

Guthrie D. and Youssef M.N. (1970). Empirical Evaluation of Some Chi-Square Tests for the Order of a Markov Chain. *Journal of the American Statistical Association* 65: 631-634.

Hardin J.W. and Hilbe J.M. (2003). Generalized Estimating Equations. Chapman and Hall: London.

Heagerty P.J. 2002. Marginalized Transition Models and Likelihood Inference for Longitudinal Categorical Data. Biometrics 58: 342-351.

Heagerty P.J. and Zeger S.L. (2000). Marginalized Multi-level Models and Likelihood Inference (with Discussion). *Statistical Science,* 15:1-26.

Health and Retirement Study, (Wave [1-7]/Year [1992-2004]) public use dataset. Produced and distributed by the University of Michigan with funding from the National Institute on Aging (grant number NIA U01AG09740). Ann Arbor, MI.

Hoel PG. (1954). A Test for Markoff Chains. *Biometrika,* 41:430-433.

Hosmer D.W. and Lemeshow S. (1989). Applied Logistic Regression. Chichester: John Wiley and Sons.

Hosmer D.W. and Lemeshow S. (2000). Applied Logistic Regression. Second Edition. John Wiley and Sons; New York.

Hu F.B., Goldberg J., Hedeker D., Flay B.R. and Pentz M.A. (1998). Comparison of Population-Averaged and Subject-Specific Approaches for Analyzing Repeated Binary Outcomes. *American Journal of Epidemiology* 147: 694-793.

Islam M.A. and Chowdhury R.I. (2006). A higher-order Markov model for analyzing covariate dependence. *Applied Mathematical Modelling* 30:477-488.

Islam M.A. and Chowdhury R.I. (2008). *Chapter 4: First and higher order transition models with covariate dependence.* In *Progress in Applied Mathematical Modeling,* F. Yang (ed), Nova Science, New York, 153-196.

Islam M.A., Chowdhury R.I., Chakraborty N. and Bari W. (2004). A Multistage Model for Maternal Morbidity During Antenatal, Delivery and Postpartum Periods. *Statistics in Medicine*, 23: 137-158.

Jones M. and Crowley J. (1992). Nonparametric tests of the Markov model for survival data, *Biometrika*, 79-3:513-522.

Kalbfleisch J.D. and Lawless J.F. (1985) The Analysis of Panel Data Under a Markov Assumption. *Journal of American Statistical Association,* 88: 863

Kalbfleisch J.D. and Prentice R.L. (2002) The Statistical Analysis of Failure Time Data. Second edition; Wiley.

Katz R.W. (1981). On Some Criteria for Estimating the Order of a Markov Chain. *Technometrics* 23: 243-249.

Kelton W.D. and Kelton C.M.L. (1984). Hypothesis Tests for Markov Process Models Estimated from Aggregate Frequency Data. *Journal of the American Statistical Association* 79: 922-928.

Korn E.L. and Whittemore A.S. (1979). Methods of Analyzing Panel Studies of Acute Health Effects of Air Pollution. *Biometrics*, 35: 795-802.

Lindsey J.K., and Lambert P. (1998). On the appropriateness of marginal models for repeated measures in clinical trials. *Statistics in Medicine*, 17: 447-469.

McCullagh P. and Nelder J.A. (1989). *Generalized Linear Models.* Second Edition. Chapman and Hall/CRC Press; London.

McCullagh P. and Nelder J.A. (2000). *Generalized, Linear and Mixed Models.* John Wiley and Sons; New York.

Miller R.G. (1963). Stationarity Equations in Continuous Time Markov Chains. *Transactions of the American Mathematical Society* 109: 35-44.

Muenz L.R. and Rubinstein L.V. (1985). Markov Models for Covariate Dependence of Binary Sequences. *Biometrics,* 41 : 91-101.

Pegram G.G.S. (1980). An autoregressive Model for Multilag Markov Chains. *Journal of Applied Probability*, 17: 350-362.

Prentice R. and Gloeckler L. (1978). Regression Analysis of Grouped Survival Data with Application to Breast Cancer Data. *Biometrics,* 34:57-67.

Raftery A.E and Tavare S. (1994). Estimating and Modeling Repeated Patterns in Higher Order Markov Chains with the Mixture Transition Distribution Model. *Appl. Statist.,* 43(1):179-199.

Raftery A.E. (1985): A Model for Higher Order Markov Chains. *Journal of Royal Statististical Society B,* 47:1528-39.

Reeves G.K. (1993). Goodness-of-Fit Tests in Two-State Processes. *Biometrika* 80: 431-442.

Regier M.H. (1968). A Two State Markov Model for Behavior Change. *Journal of American Statistical Association*, 63: 993-999.

Robinson L.D. and Jewell N.P. (1991). Covariate Adjustment. *Biometrics*, 47(1):342-343.

Sundberg R. (1986). Tests for Underlying Markovian Structure from Panel Data with Partially Aggregated States. *Biometrika* 73: 717-721.

Ten Have T.R., Landis J.R. and Hartizel J. (1996). Population-Averaged and Cluster-Specific Models for Clustered Ordinal Response Data. *Statistics in Medicine* 15: 2573-2588.

Wedderburn R.W.M. (1974). Quasi-likelihood functions, generalized linear models, and the Gauss—Newton method. *Biometrika*, 61:439-447.

Wu M. and Ware J.H. (1979). On the Use of Repeated Measurements in Regression Analysis with Dichotomous Responses. *Biometrics*, 35:513-522.

Yakowitz S.J. (1976). Small-Sample Hypothesis Tests of Markov Order, with Application to Simulated and Hydrologic Chains. *Journal of the American Statistical Association* 71: 132-136.

Young M.L., Preisser J.S., Qaqish B.F. and Wolfson M. (2007). Comparison of Subject-Specific and Population Averaged Models for Count Data from Cluster Unit Intervention Trials. *Statistical Methods in Medical Research* 16: 167-184.

ACKNOWLEDGMENTS

We would like to express our gratitude to the sponsors of the Health and Retirement Survey (HRS) data for allowing us to use the data for our research. We are also grateful to the BIRPERHT and Dr. Halida Hanum Akhter for the kind support with data as well as materials regarding the Maternal and Morbidity Survey conducted in Bangladesh.

M. Ataharul Islam
Rafiqul I Chowdhury
Shahariar Huda

May 25, 2008

Subject Index